Nanophytomedicine

Sarwar Beg • Md Abul Barkat •
Farhan Jalees Ahmad
Editors

Nanophytomedicine

Concept to Clinic

Springer

Editors
Sarwar Beg
Department of Pharmaceutics, School of Pharmaceutical Education and Research
Jamia Hamdard
New Delhi, Delhi, India

Md Abul Barkat
Department of Pharmaceutics, College of Pharmacy
University of Hafr Al-Batin
Hafr Al-Batin, Saudi Arabia

Farhan Jalees Ahmad
Department of Pharmaceutics, School of Pharmaceutical Education and Research
Jamia Hamdard
New Delhi, Delhi, India

ISBN 978-981-15-4908-3 ISBN 978-981-15-4909-0 (eBook)
https://doi.org/10.1007/978-981-15-4909-0

This Springer imprint is published by the registered company Springer Nature Singapore Pte Ltd.
The registered company address is: 152 Beach Road, #21-01/04 Gateway East, Singapore 189721, Singapore

Contents

About the Editors

Sarwar Beg is currently serving as an Assistant Professor at the Department of Pharmaceutics, School of Pharmaceutical Education and Research, Jamia Hamdard, New Delhi, India. Prior to joining Jamia Hamdard, Dr Beg was a Research Scientist at Jubilant Generics Limited, Noida, India. He has over a decade of research experience in the field of pharmaceutics, especially in the systematic development and characterization of diverse drug delivery systems employing Quality by Design paradigms like Design of Experiments and Multivariate Statistical Techniques. To date he has authored over 170 publications in various peer-reviewed journals, 45 book chapters and 12 books, with H-index of 30 and 3500 citations. Dr Beg has also participated in and presented his research work at several conferences in India, China, Bangladesh, UAE, USA and Canada, and has several best paper awards and young scientist awards to his credit.

Md Abul Barkat is currently serving as an Assistant Professor at the Department of Pharmaceutics, College of Pharmacy, University of Hafr Al Batin, Kingdom of Saudi Arabia. He holds a Master's degree in Pharmaceutical Sciences from Jamia Hamdard University, New Delhi, India, and a PhD in Pharmaceutical Sciences (Pharmaceutics) from the Integral University Lucknow, India. With more than 10 years of teaching and research experience in the field of Pharmaceutical Science, his research interests include the development and optimization of herbal/synthetic based nanostructured delivery systems, controlled release drug delivery systems, bio-enhanced drug delivery systems, nanomaterials and nanocomposites for a variety of conditions like burns, cancer etc. To date he has authored more than 30 research and review publications in various peer-reviewed journals, 6 book chapters and 7 books.

Farhan Jalees Ahmad is currently serving as a Professor at the Department of Pharmaceutics, School of Pharmaceutical Education and Research, and is Dean of School of Interdisciplinary Sciences at Jamia Hamdard, New Delhi, India. An internationally respected researcher in the area of Pharmaceutical Sciences with M. Pharma and PhD (Medicine) degrees from Jamia Hamdard, he continues to teach and leads a very productive research group, which has been extensively supported by national and international funding agencies. Prof Ahmad has 28 years of experience in research and education, with focus areas including the development, scale-up, technology transfer and launching of pharmaceutical products, both for domestic and international markets. He has published more than 300 research and review papers, 12 book chapters and 9 books.

Nanotechnology-Based Phytotherapeutics: Current Status and Challenges

Md Abul Barkat, Harshita, Sabya Sachi Das, Sarwar Beg, and Farhan J. Ahmad

Abstract

Nanotechnology has the potential to overcome numerous shortcomings associated with conventional phytotherapeutics. Notable steps have been taken towards the reinforcement of nanophytomedicine in the treatment of a variety of illnesses with high specificity, sensitivity, and efficiency. It also improves and overcomes the biopharmaceutical challenges involved with phytomedicine. Phytomedicine/herbal medicine has been practiced and used by humans for thousands of years because of their healing potential. The faith and assumption that the medicines from the herbal origin are much more reliable and safer than synthetic drugs have earned demand in recent years and managed a tremendous growth of phytopharmaceuticals. For optimized and better acceptability of herbal medicine over synthetic medicine, the equivalent robust scientific and clinical approaches should be applied. Also, there is an urgent need to develop a validated therapeutic, safety, and toxicity profile associated with herbal medicine to avoid any potential side effects. In the contemporary chapter, we endeavored to provide a short overview of application of nanotechnology approaches in the development of herbal-based remedy as well as also highlight the current situation and prospects.

M. A. Barkat · Harshita (✉)
Department of Pharmaceutics, College of Pharmacy, University of Hafr Al-Batin, Hafr Al-Batin, Saudi Arabia

S. S. Das
Department of Pharmaceutical Sciences and Technology, BIT, Mesra, Ranchi, Jharkhand, India

S. Beg · F. J. Ahmad
Department of Pharmaceutics, School of Pharmaceutical Education and Research, Jamia Hamdard, New Delhi, Delhi, India

S. Beg et al. (eds.), *Nanophytomedicine*,
https://doi.org/10.1007/978-981-15-4909-0_1

Keywords

Nanotechnology · Phytotherapeutics · Biopharmaceutical · Herbal medicine · Phytomedicine · Clinical

1 Introduction

Nanotechnology includes a set of methods and approaches which utilize materials or excipients at the nanoscale range to develop products with the novel, distinct, and enhanced physicochemical and pharmacological activities. Nanotechnology-based approaches or techniques are involved in numerous extents of knowledge to endorse improvements in the nanomedicine field, thus offer a pronounced and impending application in the healthcare system and so have provided new approaches for the enhancement of therapeutics [1, 2]. Significant attention has been focused in the past few years over the development and formulation of herbal drugs based novel drug delivery system (NDDS). Moreover, these herbal drugs conjugated novel nanocarriers should preferably accomplish two requisites. Firstly, they should deliver the therapeutically active moieties to the body at a perquisite rate and as per the rate required by the body, throughout treatment. Secondly, they should target the active moieties to the targeted sites (cells or tissues) of the body. As per the reports, conventional dosage forms such as prolonged-release dosage forms are incapable to meet these ideal conditions [3].

As per World Health Organization (WHO), the definition of herbal medicine could be justified as the practice of medicine which involves herbs, herbal ingredients, herbal formulations, and fabricated herbal extracts, which mainly comprises therapeutically active phytoconstituents isolated from the plant's parts or plant extracts, or even their combinations [4]. Generally, these herbal constituents are isolated from the plant parts including roots, stems, leaves, flowers, seeds, or the by-products (gums, resins, and many more) [5]. As per the literature and reported studies, almost 50,000 plant species have been showed to have medicinal activities [6]. About 80% of today's world's populations particularly in the developing countries, herbal medicines are still the first choice remedy for the management of a variety of illness and so it plays a vital role in today's health wellness [7–9]. Moreover, both the physicians and the patients have preferred herbal medicines or phytomedicines due to their impending therapeutic effects and less adverse effects as compared to other conventional medicines [10, 11], also the phytomedicines have been able to enhance the bioavailability [12]. Phytomedicine based therapy could be efficiently used to prevent and treat various malignancies as they are mainly focused on the naturally occurring chemical entities from various plant parts. Also as compared to chemotherapy, they have significantly enhanced the health conditions of ill people, with reduced adverse effects [13]. Unfortunately, in earlier days, due to some of the reasons such as lack of scientific rationalization and processing complications, phytomedicines were not considered for the development of novel drug delivery systems or formulations. To overcome these issues, modern phytopharmaceutical

Table 1 Glimpse of some bioactive phytoconstituents loaded nanoformulations

Novel drug delivery formulations	Bioactive phytoconstituents
Nanoparticles and microparticles	Catechins, cuscuta chinensis, camptothecin, curcumin, hypericin, triptolide, and tetrandrine
Liposomes and proliposome	Catechins, camptothecin, curcumin, silymarin, and vincristin
Solid lipid nanoparticles	Camptothecin, curcumin, tetrandrine, triptolide, and podophyllotoxin
Polymeric conjugations	Camptothecin and podophyllotoxin
Microemulsions	Triptolide, babchi oil (psoralen)
Transdermal patches and gels	Guarana extract and khellin
Solid dispersion and matrix tablets	Silymarin
Polymeric cardiac stents	Curcumin
Scaffolds	Tetrandrine

approaches could be significantly used to solve the scientific requirements of the herbal medicines in the development of novel nanocarrier systems such as microemulsion, nanoemulsion, nanoparticles, liposomes, matrix system, solid dispersion, and many more. Generally, the complication of the active constituents or moieties has made the development of a novel herbal-based drug delivery system very perplexing. It has been observed that only a limited quantity of administered doses of most of the conventional dosage forms reaches the targeted site; however, the majority of the drugs get circulated all through the body depending upon the biochemical and physicochemical properties [14, 15], which leads to low therapeutic value.

NDDS for phytomedicines involves targeted drug delivery systems, which have shown potentials to reduce the dosage frequency, enhances the solubility and absorption while decreases the elimination [16]. The nanoparticles (NPs) based drug delivery systems, in respective to herbal medicine, have been considered to be one of the most important kinds of delivery systems among all the different types of NDDS. Moreover, the NPs could be specifically utilized to target the phytomedicines to individual organs, cells, and tissues, which eventually improves the targeting ability, efficiency, and safety of the medicine. The NDDS is primarily designed and fabricated to overcome the limitations associated with traditional or conventional herbal drug delivery systems [17]. Furthermore, a list of herbal-based drug delivery approaches has been reported and is summarized in Table 1 [18].

1.1 Phytomedicine Historical Perspective

According to the reports of the ancient Babylon, plants or plant parts have been utilized as medicines some around 60,000 years ago. In Egypt and China, reported materials related to the usage of phytomedicines have been reported some around

5000 years back, whereas in Asia Minor and Greece it has been reported nearly 2500 years ago [19]. Moreover, numerous types of practices, philosophy and herbal medicinal systems or herbal-based approaches have been reported, and more interestingly each of them is significantly influenced by the region or area from which it first got evolved [20]. In China, herbal medicines persisted their development from Sheng-Nongs Herbal Book (around 3000 B.C.) to the existing updated record of medicinal plants utilized in Traditional Chinese Medicine (TCM) which involves 11,146 species of plants/herbs. Furthermore, *"The Devine Farmer's Classic of Herbalism"* is the most ancient acknowledged book describing herbal medicine, which was compiled in China (2000 years ago); abundant herbal pharmacopeias and numerous monographs on specified herbs occurs through the compiled pieces of evidence and statistics on herbs or plants [21]. In the first century A.D., Dioscorides has inscribed a very informative compilation on phytomedicines referred to as *"De Materia Medica"* that became a standard reference for the western physicians or medical practitioners for the subsequent years. During the same period, Galen of Pergamum produced 130 antidotes and phytomedicines based preparations (also recognized as galenicals) which included up to 100 herbs and other constituents [22]. However, the knowledge of traditional phytomedicine or herbal medicine was conserved by Catholic monks all through the middle ages, while its general practitioner existed outside the conventional system. Around the eighth century A.D. the physicians associated with the Arabian provinces, who performed broad research over the medicinal herbs found in the regions of Europe, India, Persia, and the far East influenced the culture and trends of the Western herbal medicine. Medicinal plants remained to be the key sources of compounds used for maintaining health until the nineteenth century [23]. For over 5000 years, Ayurveda, a truly Indian traditional system of medicine has been used in several regions of India and associated regions, founded by the ancient Hindu healers and saints. Its related compilation *materia medica* has provided an inclusive explanation of more than 1500 herbs and 10,000 formulations. The Indian government has acknowledged Ayurveda as a comprehensive healthcare system as compared to western medicine practices [24]. Also, the *Kampo* medicine, the Japanese herbal medicine (1500 years back), has been reported of having a compilation of nearly 148 formulations [25].

2 Application of Nanotechnology in Herbal Medicine

From the last few decades, nanotechnology-based herbal medicines and more precise drug delivery systems have been spreading rapidly. Furthermore, a list of herbal-based nanoformulations that has been developed is been summarized in Table 2 [3]. In the past few years, liposomes based drug delivery system has progressed from a conventional liposome to a more precise and modified liposome, according to Vladimir P. Torchilin [26]. Liposomes usually are developed in the presence of phospholipids which are generally used to alter the pharmacokinetic profile of the drugs, herbs, vitamins, and enzymes. Numerous liposome based herbal formulations have been demonstrated and established, which are summarized in

Table 2 Herbal-based Novel formulations for improved efficacy

Nano-carriers	Nano-formulations	Active pharmaceuticals	Outcomes	Biological activity	Route of delivery
Liposomes	Quercetin liposomes	Quercetin	Reduced dose, enhance penetration in blood brain barrier	Antioxidant Anticancer	Intranasal
	Liposomes encapsulated silymarin	Silymarin	Improve bioavailability	Hepatoprotective	Buccal
	Liposoma artemisia arborescens	Artemisia arborescens essential oil	Targeting of essential oils to cells, enhance penetration into, cytoplasmatic barrier	Antiviral	In vitro
	Ampelopsin liposome	Ampelopsin	Increase efficiency	Anticancer	In vitro
	Flavonoids liposomes	Quercetin and rutin	Binding of flavonoids with Hb is enhanced	Hemoglobin	In vitro
	Wogonin liposome	Wogonin	Sustained release effect	Anticancer	In vivo
	Catechins liposomes	Catechins	Increased permeation through skin	Antioxidant and chemopreventive	Transdermal
	Paclitaxel liposome	Paclitaxel	High entrapment efficiency and PH sensitive	Anticancer	In vitro
	Curcumin liposome	Curcumin	Long-circulating with high entrapment efficiency	Anticancer	In vitro
	Garlicin liposome	Garlicin	Increase efficiency	Lungs Hepatoprotective	–
	Usnea acid liposome with β-CD	Usnea acid	Increase solubility and localization with prolonged- release profile	Antimycobacterial	In vitro
	Colchicine liposome	Colchicine	Enhance skin accumulation, prolong drug release, and improve site specificity	Antigout	Topical
	Breviscapine liposomes	Breviscapine	Sustained delivery of breviscapine	Cardiovascular diseases	Intramuscular

(continued)

Table 2 (continued)

Nano-carriers	Nano-formulations	Active pharmaceuticals	Outcomes	Biological activity	Route of delivery
Nanoparticles (NPS)	Nanoparticles of Cuscuta chinensis	Flavonoids and lignans	Improve water solubility	Hepatoprotective and antioxidant effects	Oral
	Triptolide nanoparticle	Triptolide	Enhance the penetration of drugs through the stratum corneum by increased hydration	Anti-inflammatory	Topical
	Triptolide solid lipid nanoparticles	Triptolide	Decreasing the toxicity	Anti-inflammatory	Oral
	Artemisinin nanocapsule	Artemisinin	Sustained drug release	Anticancer	In vitro
	Radix salvia miltiorrhiza nanoparticles	R. salvia miltiorrhiza	Improve the bioavailability	Coronary heart diseases, angina pectoris and myocardial infarction	In vitro
	Taxel-loaded nanoparticles	Taxel	Enhance the bioavailability and sustained drug release	Anticancer	–
	Berberine-loaded nanoparticles	Berberine	Sustained drug release	Anticancer	In vitro
	Silibinin-loaded nanoparticles	Silibinin	High entrapment efficiency	Hepatoprotective	–
	Tetrandrine-loaded nanoparticles	Tetrandrine	Stability sustained drug release	Lung	In vitro
	Glycyrrhizic acid-loaded nanoparticles	Glycyrrhizic acid	Improve the bioavailability	Anti-inflammatory, antihypertensive	–
	Quercetin-loaded nanoparticles	Quercetin	Increase antioxidant activity and release of the drug 74 times higher	Antioxidant	In vitro

	Breviscapine-loaded nanoparticles	Breviscapine	Prolong the half-life and decrease RES uptake	Cardiovascular and cerebrovascular	Intra venous
	Zedoary turmeric oil nanocapsule	Zedoary turmeric oil	Increase the drug loading and stability of ZTO	Hepatoprotection anticancer and anti-bacterial	–
	Naringenin-loaded nanoparticles	Naringenin	Improved the release of NAR and improved its solubility	Hepatoprotective	Oral
	Curcuminoids solid lipid nanoparticles	Curcuminoids	Prolonged-release of the curcuminoids	Anticancer and antioxidant	In vitro
	CPT-encapsulated nanoparticles	Camptothecin	Prolonged blood circulation and high accumulation in tumors	Anticancer	In vitro
	Ginkgo biloba nanoparticles	Ginkgo biloba extract	Improving the cerebral blood flow and metabolism	Brain function activation	Oral
Phytosomes	Ginkgo biloba phytosomes	Flavonoids	Flavonoids of GBP stabilize the ROS	Cardio-protective, antioxidant activity	Subcutaneous
	Ginkgo select phytosome	Flavonoids	Inhibits lipid peroxidation (LPO), stabilize the ROS	Hepatoprotective, antioxidant	Oral
	Silybin phytosome	Flavonoids	Absorption of silybin phytosome from silybinis approximately seven times greater	Hepatoprotective, antioxidant for liver and skin	Oral
	Ginseng phytosome	Ginsenosides	Increase absorption	Nutraceutical, immunomodulator	Oral
	Green tea phytosome	Epigallocatechin	Increase absorption	Nutraceutical, systemic antioxidant, anticancer	Oral
	Grape seed phytosome	Procyanidins	The blood TRAP (total radical-trapping antioxidant parameter) were significantly elevated over the control	Systemic antioxidant, cardio-protective	Oral

(continued)

Table 2 (continued)

Nano-carriers	Nano-formulations	Active pharmaceuticals	Outcomes	Biological activity	Route of delivery
	Hawthorn Phytosome	Flavonoids	Increase therapeutic efficacy and absorption	Cardio-protective and antihypertensive	Oral
	Quercetin phytosome	Quercetin	Exerted better therapeutic efficacy	Antioxidant anticancer	Oral
	Curcumin phytosomes	Curcumin	Increase antioxidant activity and increase bioavailability	Antioxidant, anticancer	Oral
	Naringenin phytosomes	Naringenin	Prolonged duration of action	Antioxidant activity	Oral
Emulsion	Triptolide micro-emulsion	Triptolide	Enhance the penetration of drugs through the stratum corneum by increased hydration	Anti-inflammatory	Topical
	Docetaxel emulsion	Docetaxel	Improve residence time	Anticancer	Intravenous
	Berberine nanoemulsion	Berberine	Improve residence time and absorption	Anticancer	Oral
	Silybin nanoemulsion	Silybin	Sustained release formulation	Hepatoprotective	Intramuscular
	Quercetin micro-emulsion	Quercetin	Enhance penetration into stratum corneum and epidermis	Antioxidant	Topical
	Self-nanoemulsifying zedoary essential oil	Zedoary turmeric oil	Improved aqueous dispersibility, stability and oral bioavailability	Hepatoprotection anticancer and anti-bacterial	Oral
Transferosomes	Capsaicin transferosomes	Capsaicin	Increase skin penetration	Analgesic	Topical
	Colchicine transferosomes	Colchicine	Increase skin penetration	Antigout	In vitro
	Vincristine transferosomes	Vincristine	Increase entrapment efficiency and skin permeation	Anticancer	In vitro

Ethosome	Matrine ethosome	Matrine	Improve the percutaneous permeation	Anti- inflammatory	Topical
	Ammonium glycyrrhizinate ethosomes	Ammonium glycyrrhizinate	Increase of the in vitro percutaneous permeation	Anti-inflammatory	Topical
Microspheres	Rutin–alginate–Chitosan microcapsules	Rutin	Targeting into cardiovascular and cerebrovascular region	Cardiovascular and cerebrovascular diseases	In vitro
	Zedoary oil microsphere	Zedoary oil	Sustained release and higher bioavailability	Hepatoprotective	Oral
	CPT loaded microspheres	Camptothecin	Prolonged-release of camptothecin	Anticancer	Intraperitoneally and intravenously
	Quercetin microspheres	Quercetin	Significantly decreases the dose size	Anticancer	In vitro
	Cynara scolymus microspheres	Cynara scolymus extract	Controlled release of nutraceuticals	Nutritional supplement	Oral

Table 2 [3]. Due to their exceptional characteristics, liposomes are capable of enhancing the activity of the products as they significantly enhance the solubility, bioavailability, intracellular uptake, and altered pharmacokinetics as well as the biodistribution of the incorporated active moiety [27]. Also, it has been able to enhance the in vitro and in vivo stability. Liposomes based drug delivery systems can upsurge the therapeutic efficacy and safety of drugs, essentially by distributing them to their specific sites in the body and also by preserving the therapeutic levels of the drug for a prolonged period [28–30]. Numerous conventional chemotherapeutics associated with the herbal bioactive have exhibited minimum adverse effects during their passive targeting into tumor microenvironment [31]. Milk thistle (*Silybum marianum*) is one of the herbal drugs whose exceptional pharmacological profile readily provides itself for proof of clinical efficiency [3, 30]. PEG-coated liposomes (PEGylated liposomes), due to their ability of passive targeting, have enabled the substantial accumulation of the incorporated drug or herbal bioactive in the interstitial fluid at the tumor sites [32–34]. Similar to the liposomal formulation, niosomes based formulation has possessed similar structures and is comprised of nonionic surfactant-based vesicles to uphold both the hydrophilic and hydrophobic entities within the structure. Recently, Ye Jin et al. (2013) and coworkers have established a novel *Ginkgo biloba* extract loaded niosomes, which have been able to increase the bioavailability of the extract [35].

In recent years, the nanonization or size reduction of herbal medicines has fascinated many considerations and some of them are summarized in Table 2. As compared to the crude drugs, nanonization of such entities have possessed numerous advantages including enhancement in the solubility of the compound, reduction in the doses, and improvement in the absorbency of the drug or bioactive preparations [36]. Metallic nanoparticles have been widely used for pharmaceutical and diagnostic applications, as they have showed characteristic features like larger surface, specificity towards the electronic structure between the molecular and metallic states, and also processes a large number of low coordination sites. It has been noticed that the magnetic nanoparticles (MNPs) are mostly applicable in bio-separation, where the conjugation of the target biomolecules and MNPs (functionalized with specific receptors) forms complexes and could be easily fascinated in the presence of the applied magnetic field, thus provides appropriate and time-saving method as compared to conventional method like centrifugation and filtration. Further, this technique is also applicable in biosensing, drug delivery, magnetic resonance imaging, and hyperthermia [17, 37]. Solid lipid nanoparticles (SLN) such as gold and silver nanoparticles have been used for the delivery of herbal bioactive or herbal drugs [38, 39]. Nanogels act as nanocarriers for the delivery of drugs or active moieties having anticancer, antimicrobial, and anti-inflammatory activities [40]. The usage of micronized sacchachitin (mSC) nanogel for the wound healing of corneal epithelium was demonstrated in one of the studies [41].

One of the patented technology includes phytosome based systems, which has been developed by a leading manufacturer of drugs and nutraceuticals, for the incorporation of identical plant extracts or water-soluble phytoconstituents into the phospholipids to produce lipid compatible molecular complexes and so improves the absorption and bioavailability of the drug or phytoconstituents [42].

Emulsion based drug delivery systems dispense the drug or bioactive in a targeted manner due to its significant affinity towards the lymph. Additionally, the incorporated drug molecules could be designed to exhibit sustained release pattern for a prolonged period, as the drug is loaded in the inner core phase and kept off from the direct touch with tissue fluids [43]. Apart from its targeted and sustained release behavior, the emulsion system would also enhance the stability of the hydrolyzed materials, increase the penetrability of the active moieties into the skin and mucous membranes, and also reduce the response of the drugs towards the targeted cells or tissues. To date, emulsions of some herbal drugs including camptothecin, *brucea javanica* oil, coixenolide oil, and zedoary oil have been made. Zhou et al. demonstrated the effects of the elemenum loaded emulsion over the A549 cells (human lung adenocarcinoma cells) and protein expression. From the results of the study, it was noticed that elemenum emulsion specifically inhibited the growth and proliferation of the A549 cells in vitro and it also exhibited time and dose-dependent patterns [44].

Transferosomes are one of the other forms of nanocarriers that are generally applied over the skin via a non-occluded method; further, it permeates through the stratum corneum regions as a result of the osmotic or hydration force within the skin. It could also be used as drug carriers for a variety of proteins, peptides, various small biomolecules, and herbal constituents. Transferosomes significantly penetrate the lipid lamella of stratum corneum and distribute the essential drugs or nutrients locally to preserve its functions, which ensures the maintenance of the skin [45]; moreover, with this mechanism, transferosomes of capsaicin have been developed by Xiao-Ying et al. [46].

The ethosomes could effectively be used for the efficient delivery of both lipophilic and hydrophilic drugs [47]. The reports have shown that the percutaneous absorption of the ethosomal formulation of an anti-inflammatory herbal drug (*matrine*) was increased [48]. In another study, the ethosomes based formulation allowed the penetration of the antibacterial peptide into the fibrocyte more easily [49]. Furthermore, the administration of therapeutic agents through microparticulate systems is more beneficial and advantageous because microspheres can be easily ingested as well as injected. Also, they can be modified as per the desired release profiles and site-specificity of the drugs, and also even in some cases they can be designed to provide efficient organ-targeted release [50].

Liposome derived formulation, Herbasec®, has been reported for its potential uses for numerous personal healthcare applications. Similarly, based upon Phytosome® technology, several products have been designed, established, and commercialized by Indena [3], the first Italian pharmaceutical and nutraceutical based company that developed the technique to advance as well as enhance the bioavailability of plant-based isolates or extracts through complexation with specific phospholipids. The company even patented the technology with the name tag "PHYTOSOME®," Phytosomes®, or Phyto-phospholipid complexes for the efficient delivery of phytoconstituents, which are very popular nowadays. A

Phytosome®, because of its improved ability to cross the lipid-rich bio-membranes and get through the blood circulation, is found to exhibit more bioavailability as compared to simple herbal extracts [51–53].

Similar to other strategies, hydrogels (HG) have also gained significant attention because of their high aqueous contents and potentials for numerous biomedical and clinical applications [54–59]. On the other hand, herbosomes are newly established herbal formulations which have enhanced bioavailability and therapeutic actions as compared to the conventional plant extracts. The term "*Herbo*" means plant while "*some*" means cell-like. Herbosomes based formulations involve systems comprised of phospholipids and active phytoconstituents, thus efficiently enhances the bioavailability of phytoconstituents like flavonoids, phenolics, and other hydrophilic compounds [60]. Apart from other strategies, Qu et al. developed liquiritin (an active ingredient of *Glycyrrhiza uralensis*) incorporated PAMAM dendrimer-based formulation and evaluated its absorption, bioavailability and hepatoprotective, spleen protective, detoxifying activity [13].

3 Current Progress and Challenges

In recent days, pharmaceutical scientists and researchers have shown lots of interest in designing a herbal-based drug delivery system through an efficient scientific approach. *Cuscuta chinensis* is one of the commonly used traditional Chinese herbal medicine, generally used to nurture the kidney and liver. Yet, its absorption through oral administration has shown limitations, because of the poor aqueous solubility of its major constituents (flavonoids and lignans), thus the nanoparticles for the same have been developed [16, 61]. In a recent experimental study, Cucurbitacins and Curcuminoids (anticancer herbal drug) loaded polylactic acid nanoparticles were developed using a precipitation technique [16, 62]. Moreover, traditional Chinese medicine incorporated SLNs have been developed and their activity for targeted delivery, bioavailability, and efficacy were evaluated [16, 63]. Recently, nanostructured carrier systems such as polymeric nanoparticles, SLNs, liposomes, polymeric micelles, nanoemulsions, and many others have been explored for their impending ability to deliver anticancer drugs through oral route [16, 64]. Furthermore, the oral route has provided immense potentials for the delivery of chemotherapeutic agents and thus the interest has been focused on the establishment of oral chemotherapy in oncology [16, 65].

Nanosized based NDDS for bioactive or herbal drugs can potentially improve the therapeutic activity and overcome the limitations associated with the plant or plant-derived medicines. However, considerable challenges still have remained a question for the potential accomplishment of clinically feasible therapies in this field. Studies related to the novel approaches or techniques for the control of possible interactions of nanocarriers with the biological systems have signified some of the existing challenges associated with the translation of these technologies to clinical therapies. Apart from these, few other challenges include the probability of scale-up processes that could bring inventive therapeutic methods to the market and the prospect of receiving multifunctional systems to accomplish numerous biological and therapeutic necessities [66].

The prime goal behind the herbal-based nanomedicine strategies includes improvement in the pharmacokinetic behavior of active moiety, to offer sustained and controlled release behavior of the drug or bioactive, to deliver the drug or bioactive to the specific and targeted site in the body, to improve the bioavailability of the drug or bioactive, and also to decrease the adverse effects associated with the drug or bioactive [34, 67]. The progression in herbal remedies in the drug delivery system in several research institutes or laboratories is being carried out at basic and clinical trial levels. The only requirement is to design and establish better drug delivery systems for the apt delivery of herbal drugs at the targeted sites [68]. The commercial accessibility and availability of the lipid-based system are quite recent; however, medicinal plants or derived constituents have been extensively used worldwide since ancient times and are supported in lots of countries as a source of complementary medicine. There have been numerous challenges faced during the development of new phytomedicine based products including recognition of bioactive components in the herbal drugs, the stability of active constituent, safety, and an efficient pharmaceutical dosage form [66]. During the establishment and identification of biological activity, many herbal drugs have shown superior results in in vitro studies but are not always reproducible in vivo. Poor aqueous solubility or inappropriate molecular size could be one of the major reasons for the low bioavailability and absorption rate of the active moiety [69].

The development and fabrication of herbal medicinal formulations have exhibited numerous limitations, as the active drug is usually a complex mixture of active plant constituents. The recognition of a chemical marker is a vital process for quality assurance in the concluding product. This requires the establishment of an analytical method, which would be able to recognize and quantify the chemical marker present in the extract as well as in the final product [70]. In this context, lots of work have been reported in the journal *Cancer Chemotherapy and Pharmacology* [71] and in one of the study, it was reported that Meriva®'s exhibited greater bioavailability as compared to standardized curcumin extract in rats, also the hydrolytic stability and human pharmacokinetics was enhanced in Meriva®'s [72].

4 Conclusion

The thriving area of herbal medicine and nanotechnology associated with the herbal drug delivery systems needs to be explored extensively for the distinctive medical field to develop further. The novel discoveries established by the potential field of modern phytomedicine are just too appealing to be resisted by those who admire the vitality of scientific progress. Certainly, the future of phytomedicine in modern drug development looks very promising, as long as scientists keep a probing and intent mind, without prejudice towards the perception of herbal medicine.

Hence, applications and use of "herbal therapy" in the form of nanocarriers would certainly enhance its potential for the management and treatment of various acute and chronic diseases. Numerous successful examples have been noticed with the administration of nano-based research. Natural remedies are also affluent resources of beneficial compounds bearing antioxidants and constituents that could be made useful in functional foods. The amalgamation of nanotechnology in herbal based drug technology has improved the production of pharmaceuticals that would ultimately provide efficient treatment therapy for patients. The nano-based drug delivery system will deliver the phytoconstituents at the targeted disease site.

References

1. Etheridge ML, Campbell SA, Erdman AG, Haynes CL, Wolf SM, McCullough J (2013) The big picture on nanomedicine: the state of investigational and approved nanomedicine products. Nanomedicine 9:1–14
2. Barkat MA, Harshita, Rizwanullah M, Pottoo FH, Beg S, Akhter S, Ahmad FJ (2020) Therapeutic nanoemulsion: concept to delivery. Curr Pharm Des 26(11):1145–1166
3. Ajazuddin SS (2010) Applications of novel drug delivery system for herbal formulations. Fitoterapia 81:680–689
4. WHO (2000) General guidelines for methodologies on research and evaluation of traditional medicine. World Health Organization, Geneva
5. Bent S (2008) Herbal medicine in the United States: review of efficacy, safety, and regulation. J Gen Intern Med 23(6):854–859
6. Barboza GE, Cantero JJ, Núñez C, Pacciaroni A, Espinar LA (2009) Medicinal plants: a general review and a phytochemical and ethnopharmacological screening of the native argentine Flora. Kurtziana 34(7):365–374
7. Kunle OF, Egharevba HO, Ahmadu PO (2012) Standardization of herbal medicines–A review. J Biodivers Conserv 4(3):101–112
8. Okigbo RN, Anuagasi CL, Amadi JE (2009) Advances in selected medicinal and aromatic plants indigenous to Africa. J Med Plants Res 3(2):86–95
9. Kamboj VP (2000) Herbal medicine. Curr Rev 78(1):35–39
10. Goyal A, Kumar S, Nagpal M, Singh I, Arora S (2011) Potential of novel drug delivery system for herbal system for herbal drugs. Indian J Pharmac Educ Res 45:225–235
11. Thakur L, Ghodasra U, Patel N, Dabhi M (2011) Novel approaches for stability improvement in natural medicines. Pharmacogn Rev 5:48–54
12. Williamson EM (2001) Synergy and other: interactions in phytomedicines. Andreas Luch 8 (5):401–409
13. Riyaz B, Bose S, Sharma S, Khatik GL (2018) Nanotechnology-based phytopharmaceuticals in disease management: an update. Drug Invent Today 10(8):1450–1454
14. Kumar K, Rai AK (2012) Miraculous therapeutic effects of herbal drugs using novel drug delivery systems. Int Res J Pharm 3:27–33
15. Sharma AT, Mitkare SS, Moon RS (2011) Multicomponent pharmaceutical science review and research, vol 6, pp 185–187
16. Yadev D, Suri S, Choudhary AA, Sikender M, Heman, Beg MN et al (2011) Novel approach, herbal remedies and natural products in pharmaceutical science as nano drug delivery systems. Int J Pharm Tech Res 3(3):3092–3116
17. Chakraborty K, ArunShivakumar SR (2016) Nano-technology in herbal medicines: a review. Int J Herb Med 4(3):21–27
18. Musthaba SM, Baboota S, Ahmed S, Ahuja A, Ali J (2009) Status of novel drug delivery technology for phytotherapeutics. Expert Opin Drug Deliv 6(6):625–637
19. Qazi MA (2016) MolviKI. Herbal medicine: a comprehensive review. J Pharm Res 8(2):1–5

20. WHO (2005) National policy on traditional medicine and regulation of herbal medicines. World Health Organization, Geneva
21. Wachtel-Galor S, Benzie IFF (2011) Herbal medicine. In: Benzie I, Wachtel-Galor S (eds) Herbal medicine: biomolecular and clinical aspects, 2nd edn. CRC Press/Taylor & Francis, Boca Raton
22. MohamedI SA, Borhanuddin B, Fozi N (2012) The application of phytomedicine in modern drug development. Internet J Herb Plant Med 1(2):2158–2041
23. Mendonca-Filho RR (2006) Bioactive phytocompounds: new approaches in the phytosciences. In: Ahmad I, Aqil F, Owais M (eds) Modern phytomedicine. Wiley-VCH, Weinheim, pp 1–24
24. Goyal M, Sasmal D, Nagori BP (2012) Ayurveda the ancient science of healing: an insight. In: Vallisuta O, Olimat SM (eds) Drug discovery research in pharmacognosy. InTech, Rijeka, pp 1–10
25. Watanabe K, Matsuura K, Gao P, Hottenbacher L, Tokunaga H, Nishimura K et al (2011) Traditional Japanese Kampo medicine: clinical research between modernity and traditional medicine–the state of research and methodological suggestions for the future. Evid Based Complement Alternat Med 2011:513842. https://doi.org/10.1093/ecam/neq067
26. Torchilin VP (2005) Recent advances with liposomes as pharmaceutical carriers. Nat Rev Drug Discov 4:145–160
27. Patra JK, Das G, Fraceto LF, Campos EVR, Rodriguez-Torres MDP, Acosta-Torres LS et al (2018) Nano based drug delivery systems: recent developments and future prospects. J Nanobiotechnol 16(1):71
28. Abou El Wafa AA, Mursi NM, El-Shaboury KM (2003). A pharmaceutical study on certain ocular drug delivery systems. MS Thesis. Cairo Univercity, Cairo.
29. Barragan-Montero V, Winum JY, Moles JP, Juan E, Clavel C, Montero JL (2005) Synthesis and properties of isocannabinoid and cholesterol derivatized rhamnosurfactants: application to liposomal targeting of keratinocytes and skin. Eur J Med Chem 40(10):1022–1029
30. Weiss R, Fintelmann V (2000) Herbal medicine, 2nd edn. Thieme, Stuttgart
31. Yang F, Jin C, Jiang Y, Li J, Di Y, Ni Q et al (2011) Liposome based delivery systems in pancreatic cancer treatment: from bench to bedside. Cancer Treat Rev 37(8):633–642
32. Alexander A, Saraf S, Saraf S (2015) A comparative study of chitosan and poloxamer based thermosensitive hydrogel for the delivery of PEGylated melphalan conjugates. Drug Dev Ind Pharm 41(12):1954–1961. https://doi.org/10.3109/03639045.2015.1011167
33. Paolino D, Cosco D, Cilurzo F, Trapasso E, Morittu VM, Celia C et al (2012) Improved in vitro and in vivo collagen biosynthesis by asiaticoside-loaded ultradeformable vesicles. J Control Release 162(1):143–151. https://doi.org/10.1016/j.jconrel.2012.05.050
34. Alexander A, Ajazuddin, Patel RJ, Saraf S, Saraf S (2016 Nov 10) Recent expansion of pharmaceutical nanotechnologies and targeting strategies in the field of phytopharmaceuticals for the delivery of herbal extracts and bioactives. J Control Release 241:110–124. https://doi.org/10.1016/j.jconrel.2016.09.017
35. Jin Y, Wen J, Garg S, Liu D, Zhou Y, Teng L et al (2013) Development of a novel niosomal system for oral delivery of Ginkgo biloba extract. Int J Nanomedicine 8:421–430
36. Brigger I, Dubernet C, Couvreur P (2002) Nanoparticles in cancer therapy and diagnosis. Adv Drug Deliv Rev 54:631–651
37. Boris IK, Rasika Dias HV, OxanaVK VA, Pena Y, Gomez I (2014) Solubilization, dispersion and stabilization of magnetic nanoparticles in water and non aqueous solvent: recent trands. RSC Adv 4:45354–45381
38. Moltgen CV, Herdling T, Reich G (2013) A novel multivariate approach using science-based calibration for direct coating thickness determination in real-time NIR process monitoring. Eur J Pharm Biopharm 85:1056–1063
39. Karande AD, Heng PW, Liew CV (2010) In-line quantification of micronized drug and excipients in tablets by near infrared (NIR) spectroscopy: real time monitoring of tabletting process. Int J Pharm 396:63–74

40. Kaur M, Sudhakar K, Mishra V (2018) Fabrication and biomedical potential of nanogels: an overview. Int J Polym Mater Polym Biomater 68(6):1–10
41. Chen RN, Lee LW, Chen LC, Ho HO, Lui SC, Sheu MT et al (2012) Wound-healing effect of micronized sacchachitin (mSC) nanogel on corneal epithelium. Int J Nanomedicine 7:4697–4706
42. Semalty A, Semalty M, Rawat MS, Franceschi F (2010) Supramolecular phospholipids-polyphenolics interactions: the PHYTOSOME® strategy to improve the bioavailability of phytochemicals. Fitoterapia 81:306–314
43. Lu MF, Cheng YQ, Li LJ, Wu JJ (2005) Progress of study on passive targeting of drug delivery system. Mater Rev 19:108–110
44. Sarangi MK, Padhi S (2018) Novel herbal drug delivery system: an overview. Arch Med Health Sci 6:171–179
45. Benson HA (2006) Transfersomes for transdermal drug delivery. Expert Opin Drug Deliv 6 (3):727–737
46. Long XY, Luo JB, Li LR, Lin D, Rong HS, Huang WM (2006) Preparation and in vitro evaluations of topically applied capsaicin transfersomes. Zhongguo Zhong Yao Za Zhi 31 (12):981–984
47. Touitou E, Godin B, Dayan N, Weiss C, Piliponsky A, Levi-Schaffer F (2001) Intracellular delivery mediated by an ethosomal carrier. Biomaterials 22(22):3053–3059
48. Zhaowu Z, Xiaoli W, Yangdel Z, Nianfeng L (2009) Preparation of matrine ethosome, its percutaneous permeation in vitro and anti-inflammatory activity in vivo in rats. J Liposome Res 19(2):155–162
49. Zheng Y, Hou SX, Chen T, Lu Y (2006) Preparation and characterization of transfersomes of three drugs in vitro. China J Chin Mater Med 31:728–731
50. Sanli O, Karaca I, Isiklan N (2009) Preparation, characterization, and salicylic acid release behavior of chitosan/poly(vinyl alcohol) blend microspheres. J Appl Polym Sci 111 (6):2731–2740
51. Mauri PL, Simonetti P, Gardana C, Minoggio M, Morazzoni P, Bombardelli E et al (2001) Liquid chromatography/atmospheric pressure chemical ionization mass spectrometry of terpene lactones in plasma of volunteers dosed with Ginkgo biloba L. extracts. Rapid Commun Mass Spectrom 159(12):929–934
52. Kidd PM, Head K (2005) A review of the bioavailability and clinical efficacy of milk thistle phytosome: a silybin-phosphatidylcholine complex (Siliphos). Altern Med Rev 10:193–203
53. Rossi R, Basilico F, Rossoni G, Riva A, Morazzoni P, Mauri PL (2009) Liquid chromatography/atmospheric pressure chemical ionization ion trap mass spectrometry of bilobalide in plasma and brain of rats after oral administration of its phospholipidic complex. J Pharm Biomed Anal 50(2):224–227
54. Alexander A, Ajazuddin KJ, Saraf S, Saraf S (2014) Polyethylene glycol (PEG)–Poly (N-isopropyl acrylamide)(PNIPAAm) based thermosensitive injectable hydrogels for biomedical applications. Eur J Pharm Biopharm *88*(3):575–585
55. Alexander A, Khan J, Saraf S, Saraf S (2014) Formulation and evaluation of chitosan-based long-acting injectable hydrogel for PEGylated melphalan conjugate. J Pharm Pharmacol 88 (3):575–585
56. Giri TK, Verma S, Alexander A, Ajazuddin BH, Tripathy M et al (2013) Crosslinked biodegradable alginate hydrogel floating beads for stomach site specific controlled delivery of metronidazole. Farmacia 61(3):533–550
57. Lin Y, Li W, Xu J, Boulas P (2015) Development of a NIR-based blend uniformity method for a drug product containing multiple structurally similar actives by using the quality by design principles. Int J Pharm 488(1–2):120–126
58. Storme-Paris I, Clarot I, Esposito S, Chaumeil JC, Nicolas A, Brion F et al (2009) Near InfraRed Spectroscopy homogeneity evaluation of complex powder blends in a small-scale pharmaceutical preformulation process, a real-life application. Eur J Pharm Biopharm 72(1):189–198

59. Karande AD, Liew CV, Heng PW (2010) Calibration sampling paradox in near infrared spectroscopy: a case study of multi-component powder blend. Int J Pharm 395:91–97
60. Dewan N, Dasgupta D, Pandit S, Ahmed P (2016) Review on- herbosomes, a new arena for drug delivery. J Pharmacognosy Phytochem 5:104
61. Yen FL, Wu TH, Lin GT, Cham TM, Lin CC (2008) Nanoparticles formulation of Cuscuta chinensis prevents acetaminophen-induced hepatotoxicity in rats. Food Chem Toxicol 46:1771–1777
62. Zhang JF, Hou SX, Liu HL (2007) Comparison of preparing two polylactide nanoparticles loaded lipophilic anti-cancer herb drug by nanoprecipitation method. Zhongguo Zhong Yao Za Zhi 32:303–306
63. Li Y, Dong L, Jia A, Chang X, Xue H (2006) Preparation and characterization of solid lipid nanoparticles loaded traditional Chinese medicine. Int J Biol Macromol 38(3–5):296–299
64. Robinson JR (1996) Introduction: semi-solid formulations for oral drug delivery. Bull Tech Gattefosse 89:11–13
65. DeMario MD, Ratain MJ (1998) Oral chemotherapy: rationale and future directions. J Clin Oncol 16:2557–2567
66. Bonifácio BV, Silva PB, Ramos MA, Negri KM, Bauab TM, Chorilli M (2014) Nanotechnology-based drug delivery systems and herbal medicines: a review. Int J Nanomed 9:1–15. https://doi.org/10.2147/IJN.S52634
67. Onoue S, Yamada S, Chan HK (2014) Nanodrugs: pharmacokinetics and safety. Int J Nanomed. 9:1025–1037
68. Ansari SH, Islam F, Sameem M (2012) Influence of nanotechnology on herbal drugs: a review. J Adv Pharm Technol Res 3(3):142–146
69. Kesarwani K, Gupta R (2013) Bioavailability enhancer of herbal origin: an overview. Asian Pac J Trop Biomed 3(4):253–266
70. Falcão DQ, Oliveira AP, Lima BG, ACA C, AlmeidaKB STC et al (2018) Nanotechnology in 3 phytotherapy: Current challenges of lipid-based nanocarriers for the delivery of natural products. In: Lipid nanocarriers for drug targeting. Elsevier, Amsterdam, pp 139–174
71. Marczylo TH, Verschoyle RD, Cooke DN, Morazzoni P, Steward WP, Gescher AJ (2007) Cancer Chemother Pharmacol 60(2):171–177
72. Kidd PM (2009) Bioavailability and activity of phytosome complexes from botanical polyphenols: the silymarin, curcumin, green tea, and grape seed extracts. Altern Med Rev 14 (3):226–246

Nanophytomedicine Market: Global Opportunity Analysis and Industry Forecast

Rahul Shukla, Komal Thok, Imtiyaz Alam, and Raghuraj Singh

Abstract

Phytomedicines play a decisive role since prehistoric times. Now-a-days about 50% of drugs used in formulations are obtained from natural based sources. Phytomedicines demonstrate magnificent in vitro activity and less in vivo activity because of high lipophilicity, low water solubility, poor permeability, instability, high first pass metabolism, and undesirable molecular size. These biopharmaceutical issues are calculatedly responsible for poor systemic availability. India is the second largest exporter of phytomedicine across the globe. India is second to China in this ranking, with 6600 medicinal plants. India produces over 70% of the phytomedicine claim throughout the world. Indian traditional system of medicine constituted of systems such as Ayurveda, Unani, Siddha, Yoga, Homeopathy, Naturopathy. In this book chapter authors thoroughly done literature review of the current trend of phytomedicine with special emphasis on nanomedicine in India as well as global levels.

Keywords

Nanophytomedicine · Liposome · Nanomedicine · Niosomes · SLNs

1 Introduction

Phytomedicines play a decisive role since prehistoric times. Now-a-days about 50% of drugs used in formulations are obtained from natural based sources [1]. Phytomedicines are herbal medicines with both therapeutic as well as curative properties. Herbal medicines or phytomedicines are the products derived from part

R. Shukla (✉) · K. Thok · I. Alam · R. Singh
Department of Pharmaceutics, National Institute of Pharmaceutical Education and Research-Raebareli, Lucknow, Uttar Pradesh, India

S. Beg et al. (eds.), *Nanophytomedicine*,
https://doi.org/10.1007/978-981-15-4909-0_2

of plants or as a whole plant and formulated as a crude drug or in a purified form. According to WHO around 70–90% population residing in developing nation is still dependent on traditional phytomedicine for their primary medications against disease [2]. Phytomedicine has gained worldwide and regulatory acceptance due to their medicinal property and minimum side effects [3]. Phytomedicines demonstrate magnificent in vitro activity and less in vivo activity because of high lipophilicity, low water solubility, poor permeability, instability, high first pass metabolism, and undesirable molecular size. These biopharmaceutical issues are calculatedly responsible for poor systemic availability [4]. Therapeutic effectiveness of phytomedicines depends on the capability of dosage form to deliver active compound at the desired site of action in a rate controlled manner as well as in sufficient therapeutic levels to achieve desired response. Phytomedicine is one of the alternatives for conventional allopathic therapy. Safety is one of the primary concerns regarding phytomedicines. Food and Drug Administration (FDA) has reported more than 50,000 adverse effects due to herbal and other dietary supplements [5]. Phytomedicines are composed of mainly secondary metabolites and these metabolites are chemically identified and then isolated from there source plant. Active constituents isolated from source plants are highly demanding because availability of secondary metabolites in plants is very low and depends on various factors such as phytogenic species, type of soil, time of harvest and geographical area, chemotype, anatomical part used (leaves, seeds, flower, root, etc.), storage condition, sun, humidity, etc. [6]. Reports collected from all over the word described that till date humans have explored 35,000 species of plant which are presently used in various herbal treatments. According to literature of research information only 20% of over-all passes phytochemical analysis platform although 10% are able to qualify biological screening stage, remaining other plants need to identify them as potential candidate with the help of advanced and sophisticated techniques [7]. Most notable technique is hyphenated analytical technique which gives reliable fingerprint. This get bigger class of optimistic market, as it generates income of $21.7 billion annually [8].

Implementation of nanotechnology for identification, treatment, managing, and monitoring of biological system has known to be nanomedicine. Nanotechnology enables the effective delivery of poorly water soluble or lipophilic phytopharmaceuticals or herbal based medicines. Nanomedicine based tools are helpful in achieving targeted delivery and surpassing endothelial barriers. Nanotechnology is appropriate for phytomedicines that absorbs too quickly and also improves time period in which drug remains active within a body for optimal time period [9]. There are multiple advantages of nanotechnology as it shows high drug entrapment efficiency and delivering the bio actives in high drug pay load at the desired site for its therapeutic effect. Due to smaller size it increases the surface area of drug with higher dissolution concentration in the blood. Nanomedicines provides enhanced permeation and retention (EPR) effect which improves permeation through barrier due to its nano size and retention caused due to poor lymphatic drainage at the tumor sites. Manifest passive targeting without addition of any of ligand moiety decreases side effect and reduces dose strength [10, 11]. Figure 1 represents the application of nanotechnology in phytomedicines. In this book chapter authors thoroughly

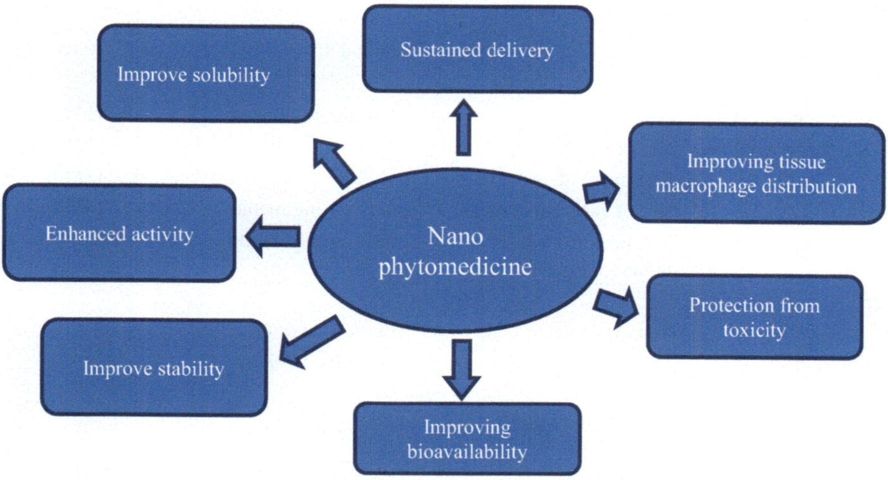

Fig. 1 Applications of nanotechnology based phytomedicines

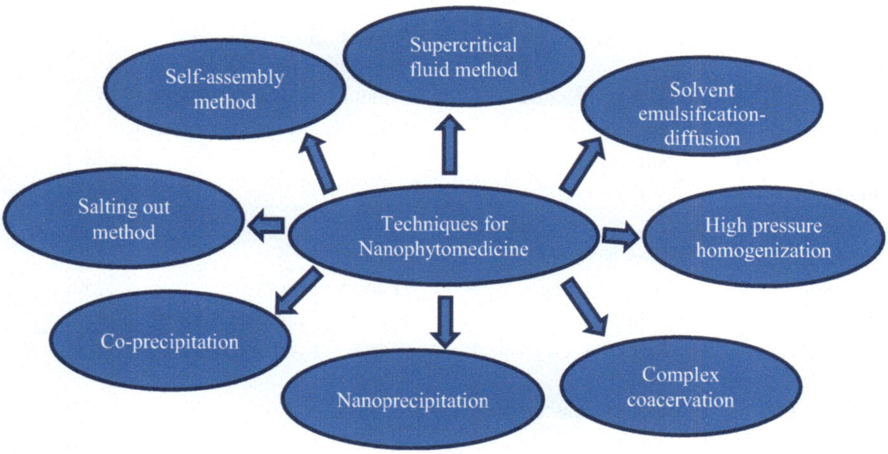

Fig. 2 Techniques used for formulating nanophytomedicine

literature the current trend of phytomedicine with special emphasis on nanomedicine in India as well as global levels (Fig. 2).

2 Classification of Nanoparticles (in Table 1)

- Labile nanoparticles: Liposomes, niosomes, polymeric micelles, solid lipid nanocarriers, nanostructured lipid carriers, nanoemulsions, etc. [12]
- Insoluble nanoparticles: titanium dioxide, silicon dioxide, fullerenes, carbon lattices, nanotubes quantum dots, etc.

Table 1 Nanoformulation of phytomedicines

Phytoconstituent	Pharmacological activity	Method	Reference
(A) Liposomes			
Avicequinone-B	Anticancer	Thin film hydration	[13]
Dihydroartemisinin	Anticancer		[14]
Psoralen	Treatment of osteoporosis		[15]
Letrozole	Breast cancer treatment	Spray drying method	[16]
Resveratrol	Treatment of muscle injury	Super critical method	[17]
(B) Niosomes			
Black tea extract	Sunscreen agent	Thin film hydration	[18]
Diallyl disulfide (garlic)	Disseminated murine candidiasis	Sonication method	[19]
Zingiber cassumunar	Anti-inflammatory		[20]
Embelin	Diabetes	Thin film hydration	[21]
Nerium oleander	Decrease cytotoxic effect		[22]
Gallic acid	Anti-skin aging		[23]
(C) Solid lipid nanoparticles			
Letrozole	Anticancer	Emulsification and solvent evaporation	[24]
Epigallocatechin	Anti-proliferative	Multiple emulsion method	[25]
Ganciclovir	Antiviral	High pressure homogenization method	[26]
Annona muricata	Anticancer		[27]
Indirubin	Anticancer		[28]
(D) Micelles			
Harmine	Liver targeting	Stirring	[29]
Galangin	Liver targeting	Dialysis method	[30]
Chrysin	Anticancer	Solvent evaporation method	[31]
Curcumin	Anti-gastroesophageal reflux	Thin film hydration method	[32]
Mangiferin	α-Glucosidase inhibitory activity		[33]
(E) Nanostructured lipid carriers			
Triptolide	Transdermal delivery	Emulsification technique	[34]
Curcuma comosa	Antioxidant activity		[35]
Silymarin	Diabetes	Emulsification and solvent evaporation	[36]
Oleuropein	Meningitis	Melt dispersion and ultrasonication	[37]
Curcumin	Anticancer	Melt emulsification	[38]
(F) Nanoparticles			
Quercetin	Antileukemic	Nanoprecipitation method	[39]
Curcumin	Anticancer		[40]

(continued)

Table 1 (continued)

Phytoconstituent	Pharmacological activity	Method	Reference
Capsaicin and gefitinib	Anticancer	Emulsification method	[41]
Pelargonium sidoides	Antibacterial		[42]
Hop flower extract	Antimicrobial		[43]
(G) Phytosomes			
Curcumin	Anticancer	Solvent evaporation technique	[44]
Morin	Antioxidant		[45]
Persimmon	Mitigating oxidative damage		[46]
Bombax ceiba	Hepatoprotective		[47]
Camellia sinensis	Antioxidant	Thin layer hydration	[48]
(H) Ethosomes			
Fisetin	Skin cancer	Thin film hydration method	[49]
Thymoquinone	Skin acne		[50]
Rosmarinic acid	Antioxidant	Mechanical dispersion method	[51]
Curcumin	Melanoma	Thin film hydration	[52]
Clove oil	Cutaneous candidiasis	Hot method	[53]

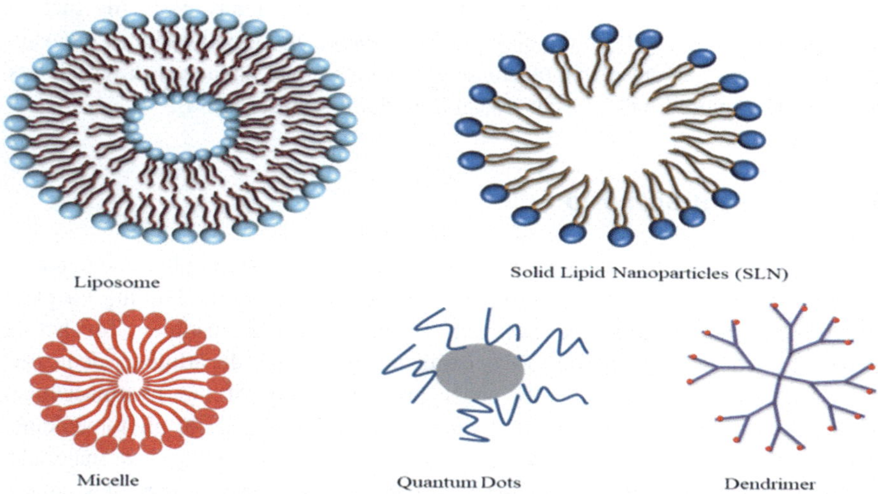

Fig. 3 Diagrammatic representation of various nanoformulations

- One-dimensional nanomaterial: nanotube and nanowire.
- Two-dimensional nanomaterials: film of monolayer self-assembled (Fig. 3).

3 Problems Encountered in Phytomedicine Development

Identification and analysis of herbal medicines is very difficult in most of the cases; this is true and most decisive in case of lower plants such as bryophytes, algae, and fungi. Several plants having wide variety of same pharmacological action, for example, more than hundreds of species have anticancer activity due to various mechanism of action [54]. In such cases, potential plants are selected for comparison of safety and efficacy and the most optimistic plant to be used for further medicinal development. Most of the time lack of good quality material is one of the major aspects, and if got active in preliminary screening then collection in bulk is an issue due to limited availability. Standardization method for medicinal plants is essential; because of presence of number of active constituent it is difficult to get efficient standardization method. Environmental changes also cause effect on quantity and quality of active constituent. Amount of active constituent varies depend on seasons, i.e. summer, winter, etc. [55].

Amount of active constituents in plants is mostly in form of secondary metabolites which are very less. Active constituents vary depending on part of plant used. Phytomedicines have one of major disadvantage due to its poor stability. Due to water content phytomedicine easily undergoes microbial degradation. Most of the phytoconstituents are thermolabile and undergo thermal degradation and stored at refrigerated conditions.

Phytoconstituents are the larger molecules and cause many problems such as solubility, permeability from the biological membrane along with reduced bioavailability. With the advent of nanotechnology various novel advanced drug delivery systems are explored with current research in pharmaceuticals [56].

4 Present Scenario

From past two decades, there is increase in awareness towards phytomedicine or herbal based medicines with government agencies supporting the funding for plant based medicine research. Development of new research institutes for herbal medicines and also upgradation of existing institutes with departments dedicated to herbal science departments is under process. Many of the pharmaceutical companies start to develop new medicines with the knowledge of phytomedicine. Demand of phytomedicine is increasing day by day for quality raw materials. Phytomedicine and plant extract estimated to rise at highest compound annual growth rate (CAGR) due to increasing demand from the pharmaceutical & dietary supplement and cosmetics industries across the world. As a manifestation of this by early 1990s top 250 pharmaceutical companies started research activities based on higher plants [57]. Phytochemicals and herbal extracts appear to have significant physiological effects on the human body. Whether they act as stimulant for enzymes, interfere with DNA replication, antioxidants, mimic hormones, destroy harmful bacteria, or bind cell walls, they could potentially curb the onset of cancer and heart diseases. These benefits, along with the natural occurrence of these extracts,

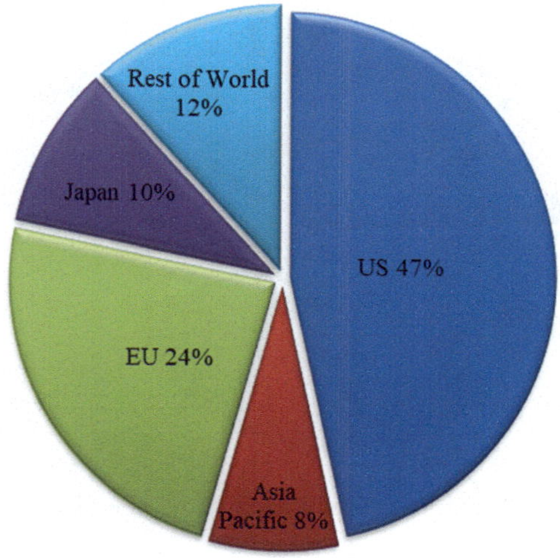

Fig. 4 Global nanomedicine market scenario of year 2025

create a growing demand for the herbal extracts across the globe. Nearby 70% urban residents of India are depend on traditional medicine for medical care [58].

India is the second largest exporter of phytomedicine across the globe. India is second to China in this ranking, with 6600 medicinal plants. India produces over 70% of the phytomedicine claim throughout the world. Indian traditional system of medicine constituted of systems such as Ayurveda, Unani, Siddha, Yoga, Homeopathy, Naturopathy, etc. [59].

The value of plant extracts market of 2019 in India is estimated at 23.7 billion USD and is expected to reach $59.4 billion by 2025, at a CAGR of 16.5% from 2019 to 2025.

The international market for plant based medicines will rise from $29.4 billion in 2017 to about $39.6 billion by 2022 with a compound annual growth rate (CAGR) of 6.1% for the period of 2017–2022.

The world marketplace of nanomedicine segment was means $53 billion in 2009 and extended to $138.8 billion in 2016. The pharmaceutical business is planned to get an aggregate selling value of nearly 334 billion USD by 2025 [60]. An anticancer product representing the largest region of nanomedicine sector in market and it is reported to be 33 billion USD in 2014. Abraxane is a human-serum albumin nanoparticle of paclitaxel and will have an approximate $967 million income in 2019 [61]. Figure 4 represents predicted value of nanomedicine market as a global scenario. Figure 5 shows nanomedicines that are in current clinical stages.

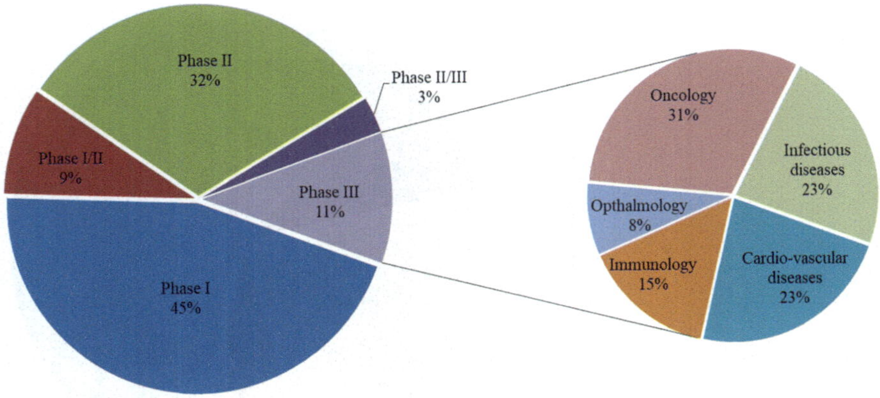

Fig. 5 Pipeline of nanomedicine product in clinical stages

Table marketed nanoformulation of phytomedicines

Brand name	Phytoconstituent	Company name
White tea liposome Herbasec®	Camellia sinensis extract	Cosmetochem
Green tea liposome Herbasec®	Camellia sinensis extract	Cosmetochem
White hibiscus liposome Herbasec®	White hibiscus extract	Cosmetochem
Aloe Vera liposome Herbasec®	Aloe vera extract	Cosmetochem
Guarana liposome Herbasec®	Guarana extract	Cosmetochem
Visnadex®	Visnadin from *Ammi Visnaga* umbel	Indena
Virtiva®	Ginkgoflavoglucosides, ginkgolides, bilobalide from *Ginkgo Biloba* leaf	Indena
Siliphos®	Sylibin from *Milk Thistle* seed	Indena
Silymarin Phytosome®	Silymarin from *Milk Thistle* seed	Indena
PA_2 Phytosome®	Proanthocyanidin A_2 from horse chestnut bark	Indena
Sericoside Phytosome®	Sericoside from *Terminalia Sericea* bark root	Indena
Meriva®	Curcuminoid from turmeric rhizome	Indena
Leucoselect® Phytosome®	Polyphenol from grape seed	Indena
Greenselect® Phytosome®	Polyphenol from green tea leaf	Indena
Ginkgo biloba dimeric flavonoid Phytosome®	Dimeric flavonoid from *Ginkgo Biloba* leaf	Indena
Ginkgo biloba terpenes Phytosome®	Ginkgolide and bilobalide from *Ginkgo Biloba* leaf	Indena
Ginselect® Phytosome®	Ginsenosides from *Panax Ginseng* rhizome	Indena
Ginkgoselect® Phytosome®	Ginkgoflavoglucosides, ginkgolides, bilobalide from *Ginkgo Biloba* leaf	Indena

(continued)

Escin β-sitosterol Phytosome®	Escin β-sitosterol from horse chestnut fruit	Indena
Crataegus Phytosome®	Vitexin-2-O-rhamnoside from Hawthorn flower	Indena
Centella Phytosome®	Triterpenes from *Centella Asiatica* leaf	Indena
18 β-glycyrrhetinic acid Phytosome®	18 β-glycyrrhetinic acid from licorice rhizome	Indena

5 Conclusion

All around the world herbal medicine recognized as a substitute for allopathic medicine. The application of herbal based drug is relatively traditional and outdated. An enormous research is conducted by various pharmaceutical companies and research professionals in the region of novel drug delivery and targeting of active phytoconstituent. A number of active phytoconstituent like alkaloids, glycosides, flavonoids, tannins, terpenoids, etc. showed increased therapeutic effect at equivalent or low doses when incorporated into novel drug delivery vesicles as compared to traditional plant extracts. Thus, there is a higher potential for development of novel drug delivery system (NDDS) for various plant derived drugs as it provides useful and cost effective drug delivery. Also the incorporation of phytomedicines into novel delivery systems has been approved by various regulatory bodies. Phytomedicine acquires substantial therapeutic activity when they deal with nanotechnology. Form the literature search it is observed that herbal medicines have excellent in vitro activity but fails to show in vivo activity due to their low water solubility, large molecule size, extensive metabolism, and instability in gastric fluid. Nanotechnology can be used to enhance the solubility as well as bioavailability of phytomedicines by particle size reduction, surface alteration, encapsulation of active phytochemical within different types of biodegradable or biocompatible polymers. Nanomaterials assist in the sustained/controlled delivery, targeted delivery as well as enhance pharmacokinetic profile of phytochemical by invading various biological barriers like blood–brain barrier. The near future research should be focused on the fabrication and development of surface engineered vesicular system and their in vivo studies to analyze the therapeutic effects.

Acknowledgment The authors acknowledge the Department of Pharmaceuticals, Ministry of Chemical and Fertilizers for providing support.NIPER R Communication no. is NIPER-R/Communication/129.

Conflict of Interest The authors declare no conflict of interest among themselves.

References

1. Naman CB, Leber CA, Gerwick WH (2017) Modern natural products drug discovery and its relevance to biodiversity conservation. Microb Resour Funct Exist Nat Appl 34:103–120

2. Srivastava A, Srivastava P, Pandey A, Khanna VK, Pant AB (2019) Phytomedicine. New look to phytomedicine. Elsevier, London, pp 625–655. https://doi.org/10.1016/B978-0-12-814619-4.00025-2
3. Sharma G, Kumar PG, Gupta RK, Block M, Alipore N (2016) Biol Bull 2(1):43–51
4. Jia L, Zhao Y (2009) Current evaluation of the millennium phytomedicine- ginseng (I): etymology, pharmacognosy, phytochemistry, market and regulations. Curr Med Chem 16 (19):2475–2484
5. Gómez JA (2009) FDA should take further actions to improve oversight and consumer understanding. Tech Rep 2009:1–71
6. Egert S, Rimbach G (2011) Which sources of flavonoids: complex diets or dietary supplements? Adv Nutr 2(1):8–14
7. Khan H (2015) Brilliant future of phytomedicines in the light of latest technological developments. J Phytopharm 4(41):58–60
8. Singh B (2016) Medicinal plants and phytomedicines. In: Herbal insecticides, repellents and biomedicines: effectiveness and commercialization. Springer, New York, pp 127–145
9. Gopi S, Amalraj A (2016) Introduction of nanotechnology in herbal drugs and nutraceutical: a review. J Nanomed Biother Discov 6(2):1–8
10. Sharma M (2016) Research and reviews: journal of pharmaceutics and nanotechnology applications of nanotechnology based dosage forms for delivery of herbal drugs. Available from https://pdfs.semanticscholar.org/7304/519a9cce5fdc18764609391 81b8c479635ae.pdf
11. Ansari SH, Islam F, Sameem M (2012) Influence of nanotechnology on herbal drugs: a review. J Adv Pharm Technol Res 3(3):142–146
12. Patra JK, Das G, Fraceto LF, Campos EVR, Rodriguez-Torres MDP, Acosta-Torres LS et al (2018) Nano based drug delivery systems: recent developments and future prospects 10 technology 1007 nanotechnology 03 chemical sciences 0306 physical chemistry (incl. structural) 03 chemical sciences 0303 macromolecular and materials chemistry 11 medical and he. J Nanobiotechnol 16(1):1–33. https://doi.org/10.1186/s12951-018-0392-8
13. Hu SCS, Su YS, Lai YC, Tseng CH, Yen FL (2019) Liposomal avicequinone-B formulations: aqueous solubility, physicochemical properties and apoptotic effects on cutaneous squamous cell carcinoma cells. Phytomedicine 58:152870. https://doi.org/10.1016/j.phymed.2019.152870
14. Li H, Li X, Shi X, Li Z, Sun Y (2019) Effects of magnetic dihydroartemisinin nano-liposome in inhibiting the proliferation of head and neck squamous cell carcinomas. Phytomedicine 56 (398):215–228
15. Li X, Garamus VM, Li N, Zhe Z, Willumeit-Römer R, Zou A (2019) Loading Psoralen into liposomes to enhance its stimulatory effect on the proliferation and differentiation of mouse calvarias osteoblasts. J Dispers Sci Technol 40(11):1531–1538
16. Maniyar M, Chakraborty A, Kokare C (2019) Formulation and evaluation of letrozole-loaded spray dried liposomes with PEs for topical application. J Liposome Res 1:1–11. https://doi.org/10.1080/08982104.2019.1634723
17. Feng Y, He Z, Mao C, Shui X, Cai L (2019) Therapeutic effects of resveratrol liposome on muscle injury in rats. Med Sci Monit 25:2377–2385
18. Yeh MI, Huang HC, Liaw JH, Huang MC, Huang KF, Hsu FL (2013) Dermal delivery by niosomes of black tea extract as a sunscreen agent. Int J Dermatol 52(2):239–245
19. Alam M, Zubair S, Farazuddin M, Ahmad E, Khan A, Zia Q et al (2013) Development, characterization and efficacy of niosomal diallyl disulfide in treatment of disseminated murine candidiasis. Nanomedicine 9(2):247–256. https://doi.org/10.1016/j.nano.2012.07.004
20. Priprem A, Janpim K, Nualkaew S, Mahakunakorn P (2016) Topical niosome gel of zingiber cassumunar Roxb. Extract for anti-inflammatory activity enhanced skin permeation and stability of compound D. AAPS PharmSciTech 17(3):631–639. https://doi.org/10.1208/s12249-015-0376-z

21. Alam MS, Ahad A, Abidin L, Aqil M, Mir SR, Mujeeb M (2018) Embelin-loaded oral niosomes ameliorate streptozotocin-induced diabetes in Wistar rats. Biomed Pharmacother 97:1514–1520. https://doi.org/10.1016/j.biopha.2017.11.073
22. Gunes A, Guler E, Un RN, Demir B, Barlas FB, Yavuz M et al (2017) Niosomes of Nerium oleander extracts: in vitro assessment of bioactive nanovesicular structures. J Drug Deliv Sci Technol 37:158–165. https://doi.org/10.1016/j.jddst.2016.12.013
23. Chaikul P, Khat-udomkiri N, Iangthanarat K, Manosroi J, Manosroi A (2019) Characteristics and in vitro anti-skin aging activity of gallic acid loaded in cationic CTAB niosome. Eur J Pharm Sci 131:39–49. https://doi.org/10.1016/j.ejps.2019.02.008
24. Yassemi A, Kashanian S, Zhaleh H (2020) Folic acid receptor-targeted solid lipid nanoparticles to enhance cytotoxicity of letrozole through induction of caspase-3 dependent- apoptosis for breast cancer treatment. Pharm Dev Technol 1:1–11. https://doi.org/10.1080/10837450.2019.1703739
25. Silva AM, Martins-Gomes C, Fangueiro JF, Andreani T, Souto EB (2019) Comparison of antiproliferative effect of epigallocatechin gallate when loaded into cationic solid lipid nanoparticles against different cell lines. Pharm Dev Technol 24(10):1243–1249. https://doi.org/10.1080/10837450.2019.1658774
26. Alelign D, Rao BVP, Obse WG (2014) Int J Acad Res 1(1):71–72
27. Sabapati M, Palei NN, Ashok AK, Molakpogu RB (2019) Solid lipid nanoparticles of Annona muricata fruit extract: formulation, optimization and in vitro cytotoxicity studies. Drug Dev Ind Pharm 45(4):577–586. https://doi.org/10.1080/03639045.2019.1569027
28. Rahiminejad A, Dinarvand R, Johari B, Nodooshan SJ, Rashti A, Rismani E et al (2019) Preparation and investigation of indirubin-loaded SLN nanoparticles and their anti-cancer effects on human glioblastoma U87MG cells. Cell Biol Int 43(1):2–11
29. Prasad Kushwaha J, Baidya D, Patil S (2019) Harmine-loaded galactosylated pluronic F68-gelucire 44/14 mixed micelles for liver targeting. Drug Dev Ind Pharm 45(8):1361–1368. https://doi.org/10.1080/03639045.2019.1620267
30. Patil S, Ujalambkar V, Rathore A, Rojatkar S, Pokharkar V (2019) Galangin loaded galactosylated pluronic F68 polymeric micelles for liver targeting. Biomed Pharmacother 112:108691. https://doi.org/10.1016/j.biopha.2019.108691
31. Baidya D, Kushwaha J, Mahadik K, Patil S (2019) Chrysin-loaded folate conjugated PF127-F68 mixed micelles with enhanced oral bioavailability and anticancer activity against human breast cancer cells. Drug Dev Ind Pharm 45(5):852–860
32. Sun Y, Li Y, Shen Y, Wang J, Tang J, Zhao Z (2019) Enhanced oral delivery and anti-gastroesophageal reflux activity of curcumin by binary mixed micelles. Drug Dev Ind Pharm 45(9):1444–1450. https://doi.org/10.1080/03639045.2019.1628041
33. Bezerra FWA, Fechine LMUD, Lopes KPS, de Sousa AF, do Nascimento GO, Amaral HH et al (2019) α-Glucosidase inhibitory activity of mangiferin-loaded F127/PEG micellar system. Mater Lett 255:126522. https://doi.org/10.1016/j.matlet.2019.126522
34. Gu Y, Tang X, Yang M, Yang D, Liu J (2019) Transdermal drug delivery of triptolide-loaded nanostructured lipid carriers: preparation, pharmacokinetic, and evaluation for rheumatoid arthritis. Int J Pharm 554:235–244. https://doi.org/10.1016/j.ijpharm.2018.11.024
35. Technology P. สารสกัดว่านชักมดลูกทีกกัเก็บในระบบตัวพาไขมันแข็งขนาดนาโนเมตรเพือเพิมการซึมผ่านผิวหนัง Curcuma comosa Roxb extract loaded nanostructured lipid carriers for enhancement of skin permeation
36. Piazzini V, Micheli L, Luceri C, D'Ambrosio M, Cinci L, Ghelardini C et al (2019) Nanostructured lipid carriers for oral delivery of silymarin: improving its absorption and in vivo efficacy in type 2 diabetes and metabolic syndrome model. Int J Pharm 572:118838. https://doi.org/10.1016/j.ijpharm.2019.118838
37. Reddy B (2019) In vitro-in vivo. Expert Opin Drug Metab Toxicol 13(2):153–166
38. Vijayakumar A, Baskaran R, Baek JH, Sundaramoorthy P, Yoo BK (2019) In vitro cytotoxicity and bioavailability of ginsenoside-modified nanostructured lipid carrier containing curcumin. AAPS PharmSciTech 20(2):1–8

39. Yin J, Hou Y, Song X, Wang P, Li Y (2019) Cholate-modified polymer-lipid hybrid nanoparticles for oral delivery of quercetin to potentiate the antileukemic effect. Int J Nanomedicine 14:4045–4057
40. Karimpour M, Feizi MAH, Mahdavi M, Krammer B, Verwanger T, Najafi F et al (2019) Development of curcumin-loaded gemini surfactant nanoparticles: synthesis, characterization and evaluation of anticancer activity against human breast cancer cell lines. Phytomedicine 57:183–190. https://doi.org/10.1016/j.phymed.2018.11.017
41. Parashar P, Tripathi CB, Arya M, Kanoujia J, Singh M, Yadav A et al (2019) A synergistic approach for management of lung carcinoma through folic acid functionalized co-therapy of capsaicin and gefitinib nanoparticles: enhanced apoptosis and metalloproteinase-9 down-regulation. Phytomedicine 53:107–123. https://doi.org/10.1016/j.phymed.2018.09.013
42. Kgatshe M, Aremu OS, Katata-seru L, Gopane R (2019) Characterization and antibacterial activity of biosynthesized silver nanoparticles using the ethanolic extract of pelargonium sidoides DC. J Nanomater 2019:3501234
43. Castro Borsari T, Germano-Costa T, Bilesky-Jose N, Fraceto LF, Lima R (2019) Antimicrobial activity of biogenic hop-based silver nanoparticles for application in sugar and alcohol industries. J Sib Fed Univ Biol 12(3):277–286
44. Tan BL, Norhaizan ME (2019) Curcumin combination chemotherapy: the implication and efficacy in cancer. Molecules 24(14):1–21
45. Manral K, Singh AK, Sah V (2019) Development and characterization of morin loaded phytosomes for its anti-oxidant activity. J Drug Deliv Ther 9(4):30–36
46. Direito R, Reis C, Roque L, Gonçalves M, Sanches-Silva A, Gaspar MM et al (2019) Phytosomes with persimmon (Diospyros kaki l.) extract: preparation and preliminary demonstration of in vivo tolerability. Pharmaceutics 11:6
47. Karole S, Karole S, Gautam GK, Gupta S (2019) Preparation and evaluation of phytosomes containing ethanolic extract of leaves of Bombax ceiba for hepatoprotective activity. Pharm Innov 8(2):22–26
48. Anwar E, Farhana N (2018) Formulation and evaluation of phytosome-loaded maltodextrin-gum Arabic microsphere system for delivery of camellia sinensis extract. J Young Pharm 10(2): s56–s62
49. Moolakkadath T, Aqil M, Ahad A, Imam SS, Praveen A, Sultana Y et al (2019) Fisetin loaded binary ethosomes for management of skin cancer by dermal application on UV exposed mice. Int J Pharm 560:78–91. https://doi.org/10.1016/j.ijpharm.2019.01.067
50. Kausar H, Mujeeb M, Ahad A, Moolakkadath T, Aqil M, Ahmad A et al (2019) Optimization of ethosomes for topical thymoquinone delivery for the treatment of skin acne. J Drug Deliv Sci Technol 49:177–187. https://doi.org/10.1016/j.jddst.2018.11.016
51. Yücel Ç, Şeker Karatoprak G, Değim İT (2019) Anti-aging formulation of rosmarinic acid-loaded ethosomes and liposomes. J Microencapsul 36(2):180–191
52. Kollipara RK, Tallapaneni V, Sanapalli BKR, Vinoth Kumar G, Karri VVSR (2019) Curcumin loaded ethosomal vesicular drug delivery system for the treatment of melanoma skin cancer. Res J Pharm Technol 12(4):1783–1792
53. Shetty S, Jose J, Kumar L, Charyulu RN (2019) Novel ethosomal gel of clove oil for the treatment of cutaneous candidiasis. J Cosmet Dermatol 18(3):862–869
54. Bonam SR, Wu YS, Tunki L, Chellian R, Halmuthur MSK, Muller S et al (2018) What has come out from phytomedicines and herbal edibles for the treatment of cancer? ChemMedChem 13(18):1854–1872
55. Heinrich M, Appendino G, Efferth T, Fürst R, Izzo AA, Kayser O et al (2020) Best practice in research – overcoming common challenges in phytopharmacological research. J Ethnopharmacol 246:112230. https://doi.org/10.1016/j.jep.2019.112230
56. Subramoniam A (2014) Present scenario, challenges and future perspectives in plant based medicine development. Ann Phytomed 3(1):31–36
57. Lahane DCC, Bhalgaon PN (2016) Research article current scenario of herbal medicine in India

58. Sen S, Chakraborty R (2017) Revival, modernization and integration of Indian traditional herbal medicine in clinical practice: importance, challenges and future. J Tradit Complement Med 7 (2):234–244
59. Nirmal SA, Pal SC, Aye T, Elachouri M, Kundu SK et al (2013) Contribution of herbal products in global market. Pharm Rev 2013:95–104
60. Bosetti R, Jones SL (2020) Cost – effectiveness of nanomedicine: estimating the real size of nano-costs. Future Med 14(2019):1367–1370
61. Barkat MA, Beg S, Pottoo FH, Ahmad FJ (2019) Nanopaclitaxel therapy: an evidence based review on the battle for next-generation formulation challenges. Nanomedicine 14 (10):1323–1341

Emergence of Nanophytomedicine in Health Care Setting

Rahul Shukla, Sanchita Kakade, Mayank Handa, and Kanchan Kohli

Abstract

Plant is considered as an important source of phytopharmaceuticals. Phytomedicine is also known as herbal medicines, herbalism, phototherapeutics, or botanical medicine. Phytomedicine is a medicinal system based on the use of plants parts or plant extracts having therapeutic efficacy. Phytoconstituents are mainly extracted from different plant parts including leaves, bark, stem, roots, rhizomes, fruits, flowers, seeds, etc. Herbal medicines are widely used in different medicinal systems such as Ayurveda, Unani, Siddha, Homeopathy, and other provincial medicinal systems. Nanomedicine based drug delivery systems for herbal drugs can possibly enhance the biological activity and overcome problems associated with phytomedicines hence more amount of drug must be delivered at targeted site for therapeutic efficacy. In this book chapter authors have thoroughly reviewed the literature about role of phytomedicine in present scenario as well as futuristic role in healthcare management.

Keywords

Phytomedicine · Nanomedicine · Phytosomes · Herbal · Phytoconstituents

R. Shukla (✉) · S. Kakade · M. Handa
Department of Pharmaceutics, National Institute of Pharmaceutical Education and Research-Raebareli, Lucknow, Uttar Pradesh, India

K. Kohli
Department of Pharmaceutics, School of Pharmaceutical Education and Research, Jamia Hamdard, New Delhi, India

S. Beg et al. (eds.), *Nanophytomedicine*,
https://doi.org/10.1007/978-981-15-4909-0_3

1 Introduction

Plant is considered as an important source of phytopharmaceuticals. Phytomedicine is also known as herbal medicines, herbalism, phototherapeutics, or botanical medicine. Phytomedicine is a medicinal system based on the use of plants parts or plant extracts having therapeutic efficacy [1]. Phytoconstituents are mainly extracted from different plant parts including leaves, bark, stem, roots, rhizomes, fruits, flowers, seeds, etc. [2, 3]. Phytoconstituents are the secondary metabolites of plant synthesized during secondary metabolism and are essential phytochemicals with potential biological activity [4]. These phytochemicals are classified as phenolics, alkaloids, terpenoids, flavonoids, tannins, glycosides, and steroids based on their chemical structure [5].

Herbal medicines are widely used in different medicinal systems such as Ayurveda, Unani, Siddha, Homeopathy, and other provincial medicinal systems. According to World Health Organization (WHO) herbal medicines are of three types: raw plant materials, processed plant materials, and medicinal herbal products [6]. In India, herbal medicines are regulated under the aegis of Ministry of Ayurveda, Yoga and Naturopathy, Unani, Siddha and Homoeopathy (AYUSH), under the Government of India. India has Ayurvedic pharmacopoeia of India and Unani pharmacopoeia of India for herbals [7].

Nanotechnology is defined as the combinatorial functionality of science and engineering carried out in the nanoscale that is 1–1000 nm. Now-a-days phytoconstituents are formulated in nanotherapeutics for enhancing their pharmacokinetic and pharmacodynamic profile [8]. Nanotechnology is the most appropriate approach to deliver an active constituent to targeted body organ. Nanophytomedicines are widely used for treatment and management of different diseases from prehistoric time [9]. Site specific targeted drug delivery and controlled or sustained drug delivery over a period of time are two main advantages which added the value to the nanomedicines over conventional dosage forms. Herbal medicines need a specific delivery system to deliver the therapeutically active components into the body hence can be formulated using the novel drug delivery system (NDDS) which is based on the nanotechnology [10]. Many novel dosage forms which incoporates phytoconstituents were developed with foremost examples like liposomes, phytosomes, niosomes, nanocapsules, nanoparticles, etc. [11]. In this book chapter authors have thoroughly reviewed the literature about role of phytomedicine in present scenario as well as futuristic role in healthcare management (Fig. 1).

2 Need of Novel Drug Delivery System for Phytoconstituents

Phytoconstituents are chemically complex in nature. Some phytoconstituents are degraded at the acidic environment of stomach while some are rapidly metabolized by liver and get converted into inactive metabolites when adminsitered orally. Hence

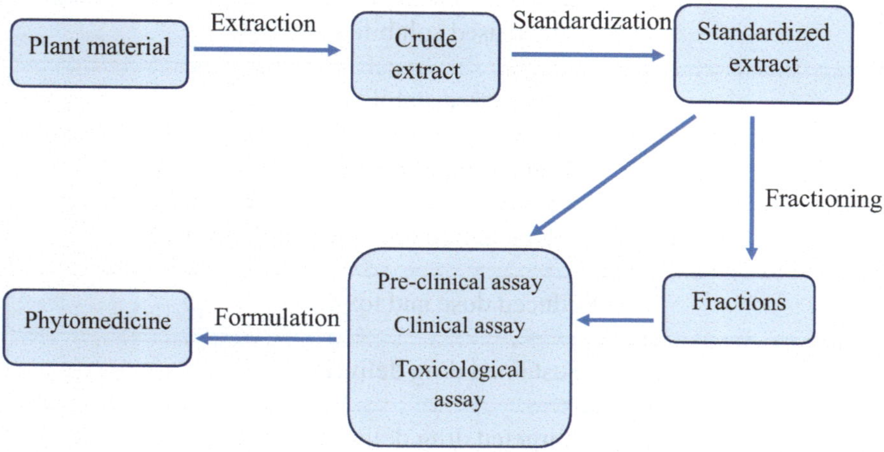

Fig. 1 Schematic representation of development of phytomedicines in various stages [12]

therapeutic plasma concentration is not achieved and phytoconstituents are unable to deliver the desired therapeutic effect. Because of all above stated hurdles, which limits the use of phytomedicines in modern medicinal system in comparison to allopathic system of medicine. All these limitations can be overcomed by using the Nanovectors to deliver the phytoconstituents [13]. Novel nanovectors are the effective carriers to deliver the phytoconstituents because of the following ways:

1. Most of the phytoconstituents are available in the form of aqueous and organic extract such as methanol, acetone, chloroform, petroleum ether, etc. which are not suitable for administration as such [14].
2. Dose of phytoconstituents is large and bioavailability is also poor hence dose reduction with simultaneous enhancement of bioavailability is necessary.
3. Site specific drug delivery can be achieved using the novel nanocarriers to treat the chronic diseases.
4. Patient compliance can be enhanced with less dose frequency.

3 Application of Novel Drug Delivery System for Phytoconstituents [15, 16]

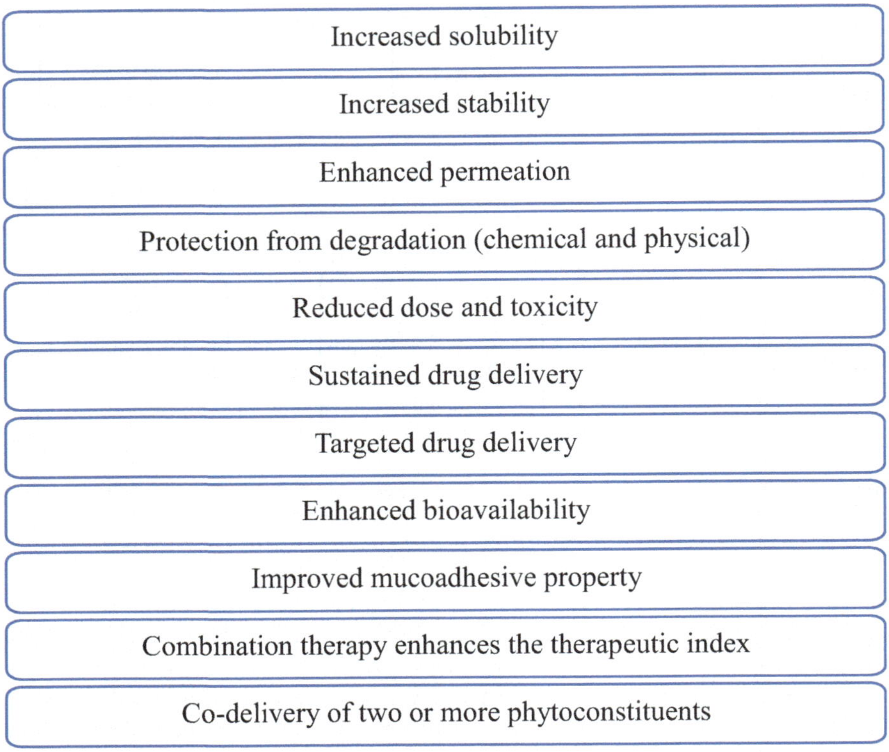

4 Novel Drug Delivery Systems for Phytoconstituents [17] (Fig. 2)

4.1 Liposomes

Liposomes are the bilayer lipid membrane with hydrophilic core. Amphiphilic lipid is used to improve efficacy and safety of drug due to its biocompatible and biodegradable property. Liposomes are used as a carrier to deliver both hydrophilic and hydrophobic drug. Depending on their size and lamellarity liposomes are classified into three groups such as small unilamellar vesicle, large unilamellar vesicle, and multilamellar vesicle [18] (Fig. 3).

4.2 Ethosomes

Ethosomes are ethanolic liposomes containing phospholipid, cholesterol, stabilizer, and high concentration of ethanol. Ethosomes are generally used for topical and transdermal drug delivery. Presence of high concentration of ethanol enhances the

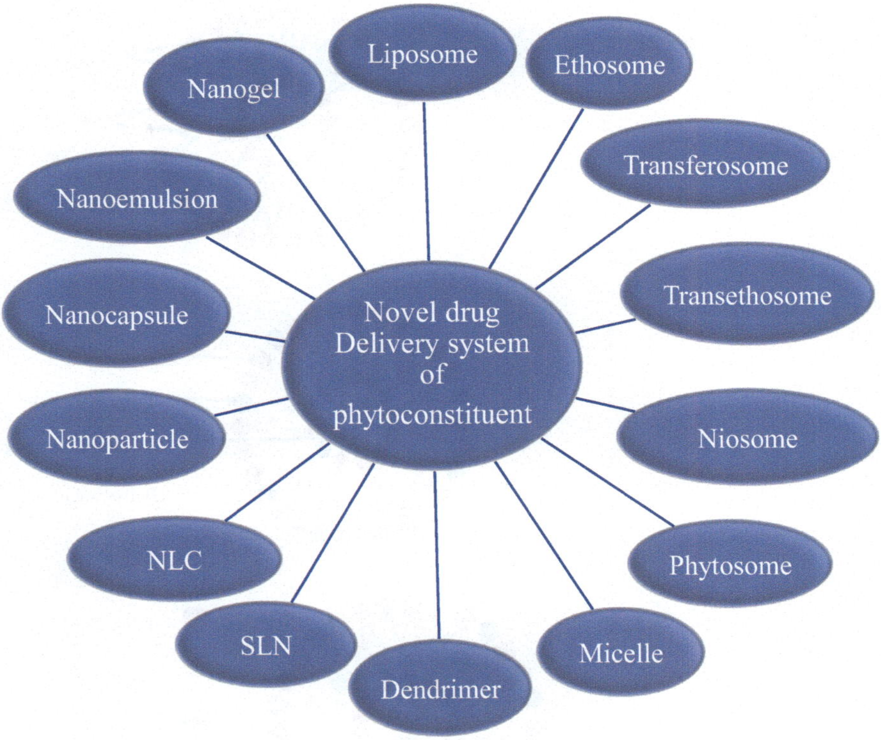

Fig. 2 Novel drug delivery systems of phytoconstituents

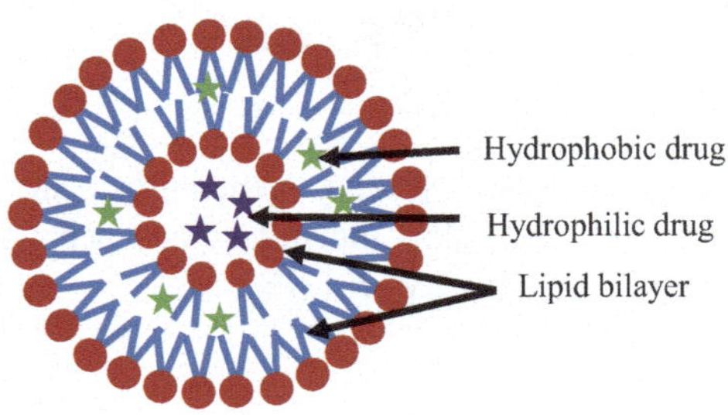

Fig. 3 Diagrammatic representation of liposomes

stability and permeability of vesicular system which helps to deliver the drug in the deeper layer of the skin [19] (Fig. 4).

4.3 Transferosomes

Transferosomes are the soft and deformable vesicular system containing the phospholipid, stabilizers, edge activator, alcohol, and high concentration of hydrophilic modulators such as organic ions. Depending on their composition transferosomes are

Fig. 4 Diagrammatic representation of ethosome

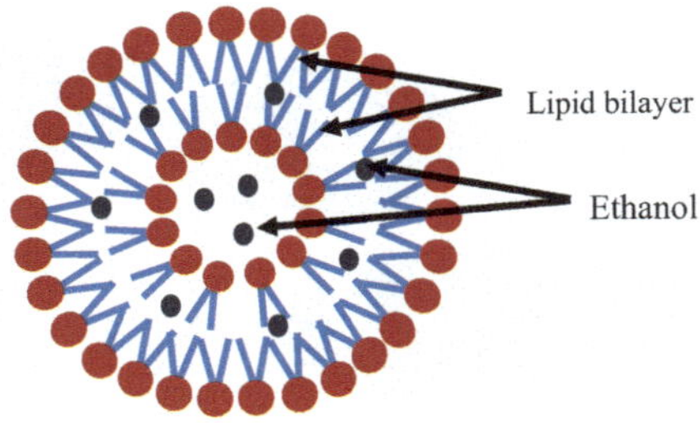

Fig. 5 Diagrammatic representation of transferosomes with various components

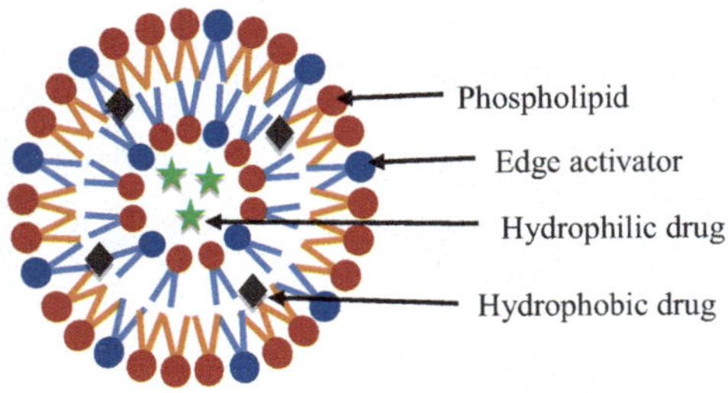

Fig. 6 Diagrammatic representation of transethosomes

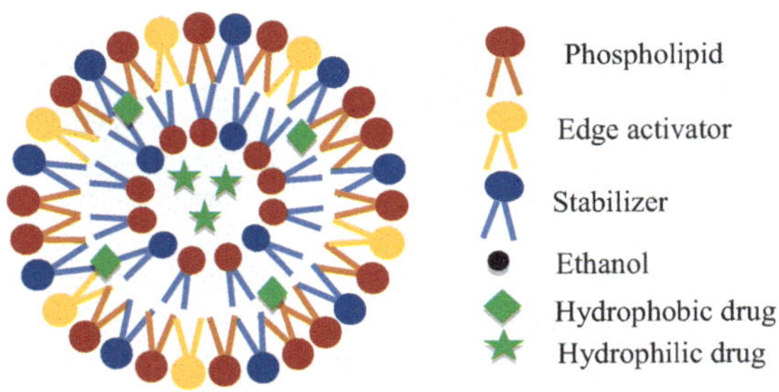

categorized as first generation, second generation, and third generation transferosomes. When applied in skin transferosomes penetrate into the deeper layer of skin without losing its integrity [20] (Fig. 5).

5 Transethosomes

Transethosomes are the modified version of classical ethosomes and transferosomes. The additional components in transethosomes are edge activators such as tween 80, span 80, sodium cholate, etc. Transethosomes are the deformable vesicles which enhance the permeation into deeper layer of skin. Transethosomes deliver the drug topically as well as systemically. Both low molecular as well as high molecular weight molecues can be incorpoarted effectively in the transethosomes and release can be in predetermined sustained manner [21] (Fig. 6).

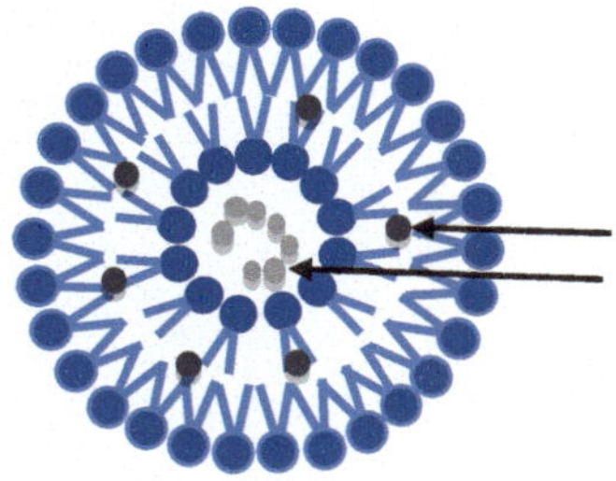

Fig. 7 Diagrammatic representation of Niosomes

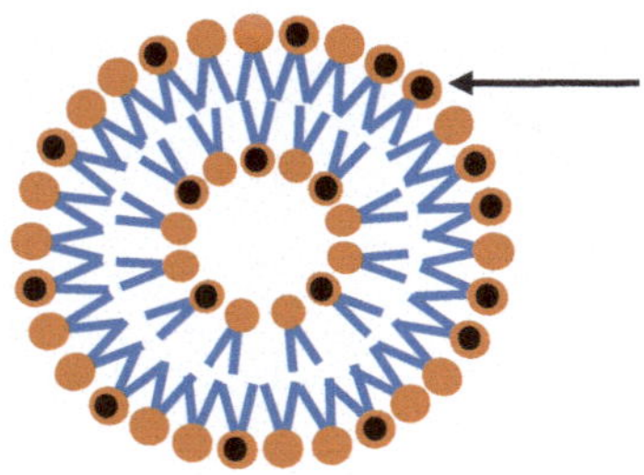

Fig. 8 Diagrammatic representation of Phytosomes

6 Niosomes

Vesicles synthesis is done from non-ionic surfactants such as tween, span, brij, etc. Which provides the solubility and stability to the phytoconstituents. The main components of niosomes are comprised of vesicle formed from non-ionic surfactants with incorporation of cholesterols which provide the mechanical rigidity to the vesicle and prevent leakage hence entrapment efficiency of niosomes. Niosomes are classified as a small unilamellar vesicle, multilamellar vesicle, and large unilamellar vesicle [22] (Fig. 7).

6.1 Phytosomes

Italian pharmaceutical and nutraceutical company Indena developed a complex of phospholipid with plant extract containing hydrophilic phytoconstituents. This molecular complex is usually formed in 1:1, 1:2 ratio and involved the H-bonding which is responsible to enhance the stability and bioavailability of phytosomes. Phytosomes are commonly term herbosomes. These herbosomes are encapsulated in different kind of formulations such as cream, solution, lotion, emulsion, gels, etc. [23] (Fig. 8).

6.2 Micelle

Micelles are self-assembled colloidal dispersion of surface-active agents or amphiphilic agents such as pluronics, polyethylene glycol (PEG), polycaprolactone (PCL),

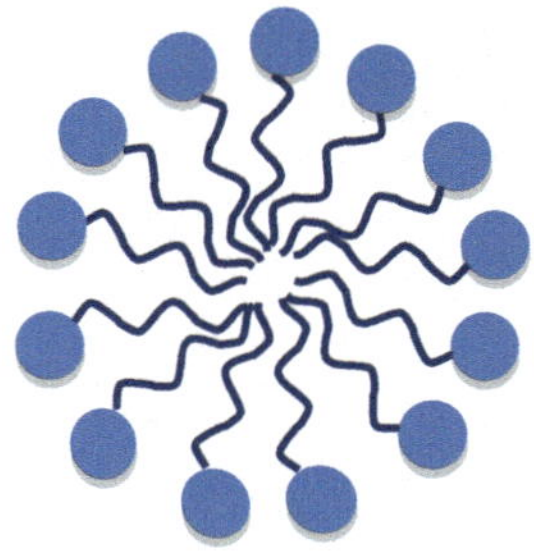

Fig. 9 Schematic representation of micelle

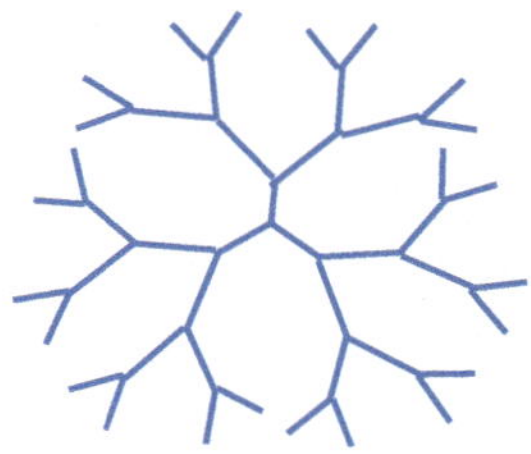

Fig. 10 Schematic representation of dendrimer

etc. Particle size of micelles usually ranges between 5 to 100 nm. These surface-active or amphiphilic agents are stable to form micelle at particular concentration and temperature, termed as critical micellar concentration and critical micellar temperature respectively. Micelles are most stable and biocompatible carrier which are also able to solubilize the hydrophobic drug molecule. Micelles also enhance the penetration of drug into interstitial compartment [24] (Fig. 9).

6.3 Dendrimer

Dendrimers are highly branched, three-dimensional structure with controllable size and shape. Number of hydrophobic and hydrophilic moieties are attached to the central core. Both drug encapsulated as well as drug-conjugated dendrimers are developed, and presently in preclinical and clinical stages. Dendrimer are used itself as carrier and therapeutic agent, with few examples of marketed formualtions. Dendrimers are commercially available and are in clinical trials or in market [25] (Fig. 10).

6.4 Solid Lipid Nanoparticles (SLNs)

Solid lipid nanoparticles are lipid based nanocarrier with particle size ranging between 10 and 1000 nm. Generally physiological lipid is preferred for SLNs due to their biocompatibility and safety. SLNs have less chances of acute and chronic toxicity as well as large scale production is possible [26]. Due to all these advantages SLNs are used as a carrier for different kind of phytoconstituents to treat chronic

Fig. 11 Schematic representation of SLN

Fig. 12 Schematic representation of NLC

diseases including lymphatic infection, cancer, neurodegenerating diseases, etc. SLNs limits its applications due to burst release upon oral administration. To overcome this problem surface modified SLN are prepared which prevent the burst release and provide targeted drug delivery [27] (Fig. 11).

6.5 Nanostructured Lipid Carriers (NLCs)

NLCs are the second generation SLNs. NLCs are composed of solid and liquid lipid which makes the imperfections in the matrix to entrap the more amount of drug. Without being affected by liquid lipid NLCs remained in solid state at room temperature by controlling the content of liquid lipid. Depending on their matrix type NLCs are classified as imperfect matrix type, amorphous matrix type, and multiple type [28] (Fig. 12).

6.6 Nanoparticles

Nanoparticles are efficiently delivering the hydrophilic as well as hydrophobic drug in submicron particle size ranging from 10 to 100 nm. Nanoparticles protect the encapsulated drug from chemical and enzymatic degradation. Recently the biodegradable polymers are preferred over non-biodegradable polymers for controlled as well as targeting drug delivery [29].

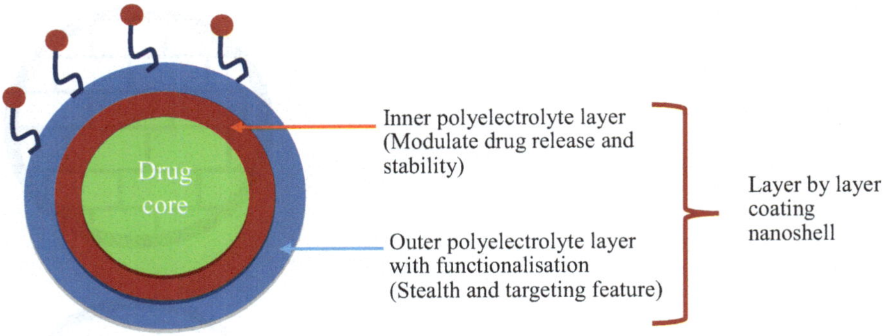

Fig. 13 Schematic representation of nanocapsule

6.7 Nanocapsules

Nanocapsules loaded phytoconstituents are formulated using the different types of biodegradable polymers such as poly (lactide) (PCL), poly-E-caprolactone) (PCL), poly (lactide-co-glycolide) (PLGA), etc. Nanocapsules are also known with other name famously as layer by layer nanoparticles [30].

. In nanocapsule the drug core is coated with oppositely charge polymers or polyelectrolytes. Phytoconstituents are easily encapsulated into the polymeric nanocapsule. Nanocapsules deliver the drug at specific site in controlled manner. Nanocapsules improve the pharmacokinetic properties and also enhance the stability and solubility of poorly water soluble phytoconstituents [16] (Fig. 13).

6.8 Nanoemulsion

Nanoemulsion is the homogeneous, kinetically stable, and isotropic system of oil, water, and blend of surfactant and cosurfactant. Droplet size of nanoemulsion ranges between 20-200 nm and size depends on the composition and homogenization method used. Phytoconstituents are both hydrophilic and lipophilic in nature. Hydrophilic phytoconstituents have poor permeability through lipid membrane which limits the therapeutic efficacy of active drug. Nanoemulsion enhances the permeability, stability, solubility of active drug which helps to enhance the bioavailability. Nanoemulsion is the promising nanocarrier for phytoconstituent [31] (Fig. 14).

6.9 Nanogel

Nanogels are the cross-linked mesh like network of polymers which can be hydrophilic, hydrophobic, and amphiphilic in nature [32]. Based on their composition gels are categorized as hydrogel and organogel with size ranging between 1 and1000 nm.

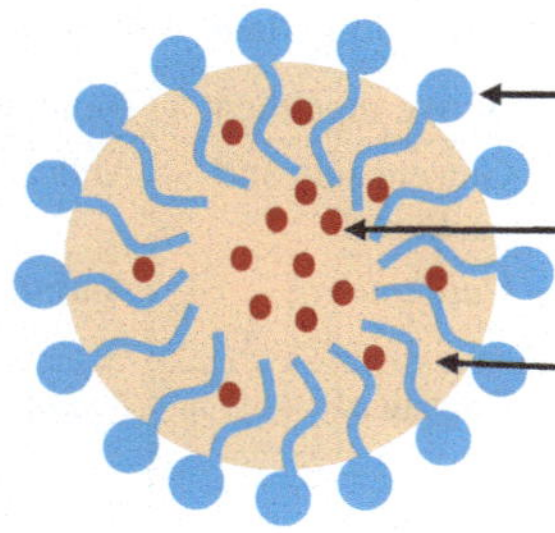

Fig. 14 Schematic representation of nanoemulsion

Nanogels are used for local and systemic action. Due to chemical modification nanogel possesses integral swelling properties which helps to release the drug [33] (Table 1).

7 Conclusion

Phytomedicine is the thrust area of research with developed nations are plunging in this field. Scientist across the globe exploring more in the area of novel drug delivery and targeting for active phytoconstituents and extracts with promising results in the investigative stage i.e. in preclinical stages [101, 102]. Revolution in the herbal medication created new opportunities and also stimulate research in the field of direct consequence to human healthcare [103]. Nanomedicine based drug delivery systems for herbal drugs can possibly enhance the biological activity and overcome problems associated with phytomedicines hence more amount of drug must be delivered at targeted site for therapeutic efficacy [104]. Development of novel drug delivery system for esteemed herbal drugs provides competent and economical drug delivery. Also, the trend of incorporating NDDS for herbal drugs has also been implemented at industrial scale [14]. Phytoconstituents are less pricey, easily available, and non-toxic compared to their synthetic analogues. Hence, in future, there is going to be continued interest in the phytomedicines to have superior materials for drug delivery systems [15].

Acknowledgments The authors acknowledge the Department of Pharmaceuticals, Ministry of Chemical and Fertilizers for providing the support. NIPER R Communication no. is NIPER-R/Communication/128.

Table 1 Different kind of formulations of phytoconstituents

Delivery system	Active ingredient	Biological activity	Rationale	Reference
Liposome	Ginsenoside and paclitaxel	Anticancer	Exert synergistic effect	[34]
	Curcumin	Anti-inflammatory	Enhanced solubility	[35]
	Capsaicin	Analgesic activity	Improved oral bioavailability	[36]
	Fisetin	Anticancer	Improved bioavailability and antitumor efficacy	[36]
	Epigallocatechin gallate	Anticancer	Enhances the stability of drug	[37]
Ethosome	Curcumin	Antioxidant, anti-inflammatory	Drug deposited into deeper layer of skin	[38]
	Tetrandrine	Anti-inflammatory	Enhanced topical drug delivery and therapeutic efficacy	[39]
	Sophoridine, matrine, sophocarpine, lemannine	Antiendotoxic, anticancer, and anti-inflammatory	Drug delivered into the deeper layer of skin	[40]
	Matrine	Anti-inflammatory	Increased percutaneous permeation	[41]
	Ammonium Glycyrrhizinate, methyl-nicotinate	Anti-inflammatory	Increased percutaneous permeation and anti-inflammatory activity	[42]
Transferosome	Quercetin	Antioxidant	Prolong drug delivery	[43]
	Resveratrol	Antioxidant	Improved the solubility, stability and safety.	[44]
	Capsaicin	Anti-inflammatory	Enhanced antiarthritic efficacy	[45]
	Curcumin	Antioxidant, antiaging, antiwrinkle	Enhanced skin permeation	[46]
	Emodin	Anti-obesity	Enhanced therapeutic efficacy	[47]
Transethosomes	Sinomenine hydrochloride	Anti-inflammatory	Enhanced transdermal permeability and drug deposition	[48]
	Colchicine	Anti-gout	Enhanced skin permeation of drug	[49]

(continued)

Table 1 (continued)

Delivery system	Active ingredient	Biological activity	Rationale	Reference
	Epigallocatechin gallate	Antioxidant	Increased skin penetration	[50]
	Paeonol	Anti-allergic, anti-inflammatory, and analgesic	Improved in vitro drug delivery and bioavailability	[51]
	Capsaicin	Analgesic, antioxidant, anticancer, and anti-obesity	Increased skin penetration	[52]
Niosomes	Lawsone	Antitumor	Improved therapeutic efficacy	[53]
	Borreline, β-sitosterol, Ursolic acid, and Isorhmnatin	Antituberculosis activity	Prolong drug release	[54]
	Lycopene	Anticancer	Increased entrapment efficiency	[55]
	Rosmarinic acid	Antibacterial and anti-inflammatory	Deliver drug into deeper layer of skin	[56]
	Gymnemic acid	Antihyperglycemic	Improved therapeutic efficacy	[57]
Phytosomes	Ethanolic extract of *Centella asiatica*	Anti-inflammatory	Increased bioavailability	[58]
	Silybin	Anticonvulsant	Increased bioavailability	[59]
	Silymarin	Hepatoprotective	Improved bioavailability	[60]
	Berberine	Hypoglycemic activity	Improved oral bioavailability and hypoglycemic efficiency	[61]
	Boswellic acid	Anti-inflammatory	Improved systemic bioavailability and tissue distribution	[62]
Micelle	Quercetin and alantolactone	Anticancer	Synergistic effect improved therapeutic efficiency	[63]
	Berberine and Diosmin	Anticancer	Dual targeted drug delivery	[64]
	Curcumin	Anti-Alzheimer	Enhanced bioavailability and brain uptake	[65]
	Puerarin	Antiinebriation	Enhanced solubility	[66]

(continued)

Table 1 (continued)

Delivery system	Active ingredient	Biological activity	Rationale	Reference
	β-Lapachone and paclitaxel	Anticancer	Biological synergy increases drug Delivery efficiency	[67]
Dendrimer	Gallic acid	Hepatoprotective	Increased bioavailability and therapeutic efficacy	[68]
	Curcumin	Anticancer	Enhanced therapeutic efficacy	[69]
	Paclitaxel	Anticancer	Efficient targeted drug delivery	[70]
SLN	Carvacrol and astaxanthin	Antioxidant and anti-biofilm	Improved water solubility and dispersibility	[71]
	Curcumin	Anti-inflammatory, antioxidant, anti-amyloidogenic effects	Enhanced oral bioavailability	[72]
	Ellagitannins	Antioxidant, anticancer	Enhanced anticancer efficacy	[73]
	Thymoquinone	Hepatoprotective	Improved oral delivery	[74]
	Emodin	Antitumor	Enhanced cytotoxicity	[75]
NLC	Capsaicin	Anti-inflammatory	Enhanced therapeutic efficacy	[76]
	Thymoquinone	Anticancer	Improved bioavailability and cytotoxicity	[77]
	Docetaxel and nicotinamide	Anticancer	Enhanced bioavailability	[78]
	β-Elemene	Antitumor	Enhanced bioavailability and antitumor efficacy	[79]
	Quercetin	Anti-inflammatory and antioxidant	Enhanced skin penetration	[80]
Nanoparticle	Paclitaxel	Anticancer	Targeted drug delivery	[81]
	Curcumin and Rutin	Anticancer	Increased bioavailability	[82]
	Quercetin	Anticancer	Sustained and controlled drug release	[83]
	Silymarin	Antioxidant	Increased solubility and antioxidant activity	[84]
	Andrographolide	Antitumor	Improved biopharmaceutical properties of active agent	[85]

(continued)

Table 1 (continued)

Delivery system	Active ingredient	Biological activity	Rationale	Reference
Nanocapsule	Hydrocotyle sibthorpioides and clove oil	Antibacterial	Improved efficacy and targeted drug release	[86]
	Catechin	Antioxidant	Enhanced bioavailability and antioxidant activity	[87]
	Paeonol	Antibacterial	Improved bioavailability and pharmacological properties	[88]
	Tetrahydrofuran neolignan grandisin	Anticancer	Enhanced therapeutic efficacy	[89]
	Artemisinin	Anticancer	Controlled drug release	[90]
Nanoemulsion	Anthocyanin	Antioxidant, anti-inflammatory, and anticancer	Improved stability and bioavailability	[91]
	Ginger oil	Antibacterial activity	Increased bioavailability and efficacy	[92]
	Polyphenon 60 (P60) and cranberry (CRB)	Antibacterial activity	Enhanced therapeutic efficacy	[93]
	Lime oil	Antibacterial activity	Enhanced therapeutic efficacy	[94]
	Lapachol	Antibacterial, antifungal, anti-inflammatory,	Enhanced bioavailability	[95]
Nanogel	Apigenin	Anticancer	Increased concentration and exposure time	[96]
	Mentha piperita oil	Anti-biofilm activity	Enhanced biological activity	[97]
	Acitretin and aloe-emodin	Anti-psoriatic	Enhanced biological activity	[98]
	Cuminum cyminum oil	Antimicrobial	Improved therapeutic efficacy	[99]
	Thyme oil	Antimicrobial	Increased half-life and antimicrobial properties	[100]

Conflict of Interest The authors declare no conflict of interest among themselves.

References

1. Lokesh BS, Kumar D, Handa M, Shukla R (2019) History of flavors associated with functional foods and nutraceuticals. In: Flavor development for functional foods and nutraceuticals. CRC Press, Boca Raton
2. Alexander A, Patel RJ, Saraf S, Saraf S (2016) Recent expansion of pharmaceutical nanotechnologies and targeting strategies in the field of phytopharmaceuticals for the delivery of herbal extracts and bioactives. J Control Release [Internet]. https://doi.org/10.1016/j.jconrel.2016.09.017
3. Yalavarthi C, Thiruvengadarajan VS (2013) A review on identification strategy of phyto constituents present in herbal plants. Int J Res Pharm Sci 4:123–140
4. Sayed N, Khurana A, Godugu C (2019) Pharmaceutical perspective on the translational hurdles of phytoconstituents and strategies to overcome. J Drug Deliv Sci Technol [Internet] 53:101201. https://doi.org/10.1016/j.jddst.2019.101201
5. Ambwani S, Ambwani T, Malik YS (2018) Current knowledge on nanodelivery systems and their beneficial applications in enhancing the efficacy of herbal drugs. J Exp Biol Agric Sci 6:87–107
6. Rodr J (2018) Pharmacological activities of phytomedicines: a challenge horizon for rational knowledge. Challenges 9:1–8
7. Yevale R, Khan N, Kalamkar P (2018) Overview on "Regulations of herbal medicine". J Pharm Phytochem 7:61–63
8. Lekshmi NG, Chandran SM, Sundar WA (2018) Role of nanotechnology in herbal medicine. Indo Am J Pharm Sci 5(11):12052–12062
9. Simona AD, Florina A, Evelyne O, Maria-corina S (2017) Nanoscale delivery systems: actual and potential applications in the natural products industry. Curr Pharm Design *23* (17):2414–2421
10. Saraf S (2010) Fitoterapia applications of novel drug delivery system for herbal formulations. Fitoterapia [Internet] 81(7):680–689. https://doi.org/10.1016/j.fitote.2010.05.001
11. Review E, Kulkarni GT (2011) Herbal drug delivery systems: an emerging area in herbal drug research. J Chronother Drug Deliv 2(3):113–119
12. Review P (2014) Nanotechnology: an effective tool for enhancing bioavailability and bioactivity of phytomedicine. Asian Pac J Trop Biomed 4(Suppl 1):S1–S7
13. Ansari SH, Islam F (2012) Influence of nanotechnology on herbal drugs: a review. J Adv Pharm Technol Res 3(3):142–147
14. Ajaz A, Siddique S, Science S, Technology A, Alexander A, Yadav P (2019) Nanomedicines: Challenges and perspectives for future nanotechnology in the healthcare system. Angew Chem Int Ed Engl. 48(5):872–897
15. Afrin S, Jahan I, Hasan AHMN, Deepa KN (2018) Novel approaches of herbal drug delivery. J Pharm Res Int 21(5):1–11
16. Botlagunta M (2014) Nanophytomedicine and drug formulations. Int J Nanotechnol Appl 4:1–8
17. Dhiman A, Nanda A, Ahmad S (2012) Novel herbal drug delivery system (NHDDS): the need of hour 2. Types of Novel Herbal Drug Delivery Systems 49:171–175
18. Sachan AK, Gupta A (2015) A review on nanotized herbal drugs. J Pharm Sci Res *6*(3):961
19. Pawar P, Kalamkar R, Jain A, Amberkar S (2015) Ethosomes: a novel tool for herbal drug delivery. IJPPR Human 3:191–202
20. Pradeep C, Kumar TB (2018) Review article 8(3):162–168
21. Ascenso A, Raposo S, Batista C, Cardoso P, Mendes T, Praça FG et al (2015) Development, characterization, and skin delivery studies of related ultradeformable vesicles: Transfersomes, ethosomes, and transethosomes. Int J Nanomedicine 10:5837–5851

22. Gharbavi M, Amani J, Kheiri-manjili H, Danafar H, Sharafi A (2018) Niosome: a promising nanocarrier for natural drug delivery through blood-brain barrier. Adv Pharm Pharm Sci 2018:15
23. Prajkta C, Vinal P, Vineeta J (2017) A review on therapeutic applications of phytosomes. J Drug Deliv Therap 7(5):17–21
24. Conte R, De Luca I, De Luise A, Petillo O, Calarco A, Peluso G (2016) New therapeutic potentials of nanosized phytomedicine. J Nanosci Nanotechnol. 16(8):8176–8187
25. Mignani S, Zablocka M, Tomas H, Caminade A (2018) Chem Soc rev and analogues as anti-cancer agents. Chem Soc Rev [Internet] 47:514–532. https://doi.org/10.1039/c7cs00550d
26. Shevalkar G, Vavia P (2019) Solidified nanostructured lipid carrier (S-NLC) for enhancing the oral bioavailability of ezetimibe. J Drug Deliv Sci Technol [Internet] 53:101211. https://doi.org/10.1016/j.jddst.2019.101211
27. Takarkhede SN, Kale MK (2017) A review on use of herbal drugs for solid lipid nanoparticles. Int J Innov Sci Res Technol 2(10):116–121
28. Jaiswal P, Gidwani B, Vyas A (2014) Nanostructured lipid carriers and their current application in targeted drug delivery. Artif Cells Nanomed Biotechnol 44(1):1–14
29. Akash B, Ravindranath S (2019) A review on novel drug delivery system : a recent trend. J Drug Deliv Therap 9(3):517–521
30. Santos AC, Pattekari P, Jesus S, Veiga F, Lvov Y (2015) Sonication-assisted layer-by-layer assembly for low solubility drug nanoformulation. ACS Appl Mater Interfaces 7:11972
31. Harwansh RK, Deshmukh R, Rahman A (2019) Nanoemulsion : promising nanocarrier system for delivery of herbal bioactives. J Drug Deliv Sci Technol [Internet] 51:224–233. https://doi.org/10.1016/j.jddst.2019.03.006
32. Baskar V, SM I, Subramani A, Ali J (2018) Historic review on modern herbal nanogel formulation and delivery methods. Int J Pharm Pharm Sci 10(10). https://doi.org/10.22159/ijpps.2018v10i10.23071
33. Lai W, Rogach AL (2017) Hydrogel-based materials for delivery of herbal medicines. ACS Appl Mater Interfaces 9:11309
34. Hong C, Wang D, Liang J, Guo Y, Zhu Y, Xia J et al (2019) Novel ginsenoside-based multifunctional liposomal delivery system for combination therapy of gastric cancer. Theranostics 9(15):4437–4449
35. Madheswaran T, Kumar P, Pillay V, Hsu A, Hansbro N, Bebawy M et al (2018) Assessing the potential of liposomes loaded with curcumin as a therapeutic intervention in asthma. Colloids Surf B Biointerfaces [Internet]. https://doi.org/10.1016/j.colsurfb.2018.08.027
36. Seguin J, Brullé L, Boyer R, Lu YM, Romano MR, Touil YS et al (2013) Liposomal encapsulation of the natural flavonoid fisetin improves bioavailability and antitumor efficacy. Int J Pharm [Internet] 444(1–2):146–154. https://doi.org/10.1016/j.ijpharm.2013.01.050
37. Fang J, Lee W, Shen S (2006) J Dermatol Sci 42:101
38. Jeswani G (2015) Drug delivery research topical delivery of curcuma longa extract loaded nanosized ethosomes to combat facial wrinkles. J Pharm Drug Deliv Res 3(1). https://doi.org/10.4172/2325-9604.1000118
39. Fan C, Li X, Zhou Y, Zhao Y, Ma S, Li W et al (2013) Enhanced Topical Delivery of Tetrandrine by Ethosomes for Treatment of Arthritis. Biomed Res Int 2013:161943
40. Zhou Y, Wei Y, Liu H, Zhang G, Wu X (2010) Preparation and In vitro Evaluation of Ethosomal Total Alkaloids of Sophora alopecuroides Loaded by a Transmembrane pH-Gradient Method. AAPS Pharm SciTech 11(3):1350–1358
41. Zhaowu Z, Xiaoli W, Yangde Z, Nianfeng L (2009) Preparation of matrine ethosome, its percutaneous permeation in vitro and anti-inflammatory activity in vivo in rats. J Liposome Res 19(87):155–162
42. Paolino D, Lucania G, Mardente D, Alhaique F, Fresta M (2005) Ethosomes for skin delivery of ammonium glycyrrhizinate : In vitro percutaneous permeation through human skin and in vivo anti-inflammatory activity on human volunteers. J Control Release 106:99–110

43. Nangare S, Dhananjay B, Shitole M (2019) Development of novel freeze-dried mulberry leaves extract-based transfersomal gel. Turk J Pharm Sci. https://doi.org/10.4274/tjps.98624
44. Wu P, Li Y, Kuo Y, Tsai SJ, Lin C (2019) Preparation and Evaluation of Novel Transfersomes Combined with the Natural Antioxidant Resveratrol. Molecules 24:1–12
45. Sarwa KK, Mazumder B, Rudrapal M, Verma VK (2014) Potential of capsaicin-loaded transfersomes in arthritic rats. Drug Deliv 7544:1–9
46. Saraf S, Jeswani G, Kaur CD, Saraf S (2011) Development of novel herbal cosmetic cream with Curcuma longa extract loaded transfersomes for antiwrinkle effect. Afr J Pharm Pharmacol 5:1054–1062
47. Lu K, Xie S, Han S, Zhang J, Chang X, Chao J et al (2014) Preparation of a nano emodin transfersome and study on its anti-obesity mechanism in adipose tissue of diet-induced obese rats. J Transl Med 12:1–14
48. Song H, Li H, Meng Y, Zhang Y, Zhang N, Zheng W (2019) Enhanced transdermal permeability and drug deposition of rheumatoid arthritis via sinomenine hydrochloride-loaded antioxidant surface transethosome. Int J Nanomedicine 14:3177–3188
49. Karim NA (2018) Transethosomal gels as carriers for the transdermal delivery of colchicine: statistical optimization, characterization, and ex vivo evaluation. Drug Des Devel Ther 12:795–813
50. Anwar E, Ramadon D, Ardi GD (2018) Novel transethosome containing green tea (CAMELLIA sinensis L. kuntze) leaf extract for enhanced skin delivery of epigallocatechin gallate: formulation and in vitro penetration test. Int J Appl Pharm 10(1):1–4
51. Chen ZX, Li B, Liu T, Wang X, Zhu Y, Wang L et al (2017) Evaluation of paeonol-loaded transethosomes as transdermal delivery carriers. Eur J Pharm Sci 99:240–245
52. Azizah N, Sagita E, Iskandarsyah I (2017) In vitro penetration tests of transethosome gel preparations containing capsaicin. Int J Appl Pharm 9:5–8
53. Barani M, Mirzaei M, Torkzadeh-mahani M, Nematollahi MH (2018) Lawsone-loaded Niosome and its antitumor activity in MCF-7 breast cancer cell line: a nano-herbal treatment for cancer. Daru 26(1):11–17
54. Activity A (2017) Development of novel nano niosomes as drug delivery system of spermacoce hispida extract and in vitro antituberculosis activity. Curr Nanomat 2:17–23
55. Activity A (2016) Niosomes and assessment of its journal of bioequivalence & bioavailability novel encapsulation of lycopene inosomes and assessment of its anti- cancer activity. J Bioequiv Bioavail. https://doi.org/10.4172/jbb.1000300
56. Budhiraja A, Dhingra G (2014) Development and characterization of a novel antiacne niosomal gel of rosmarinic acid. Drug Deliv 7544:1–8
57. Kamble B, Talreja S, Pathak D (2013) Development and biological evaluation of Gymnema sylvestre extract-loaded nonionic surfactant-based niosomes. Nanomedicine 8(8):1295–1305
58. Park JH, Jang JS, Kim KC, Hong JT (2018) Anti-inflammatory effect of centella asiatica phytosome in a mouse model of phthalic anhydride-induced atopic dermatitis. Phytomedicine [Internet]. https://doi.org/10.1016/j.phymed.2018.04.013
59. Zaidi SMA, Pathan SA, Jamil SS, Ahmad FJ, Khar RK, Singh S (2016) Enhanced neurobehavioral effects of Jadwar (Delphinium denudatum) aqueous fraction by implying nanotechnology based approach. Archiv Neurol Neurosurgery 1(1):1–6
60. Maryana W, Rachmawati H, Mudhakir D (2016) Symposium on flexible organic electronics. Formation of phytosome containing silymarin using thin layer-hydration technique aimed for oral delivery. Mater Today Proc 3:855–866
61. Yu F, Li Y, Chen Q, He Y, Wang H, Yang L et al (2016) Monodisperse microparticles loaded with the self-assembled phytosomes for improving oral bioavailability and enhancing hypoglycemic efficiency. Eur J Pharm Biopharm [Internet]. https://doi.org/10.1016/j.ejpb.2016.03.019
62. Hüsch J, Bohnet J, Fricker G, Skarke C, Artaria C, Appendino G et al (2013) Fitoterapia Enhanced absorption of boswellic acids by a lecithin delivery form (Phytosome ®) of Boswellia extract. Fitoterapia [Internet] 84:89–98. https://doi.org/10.1016/j.fitote.2012.10.002

63. Zhang J, Shen L, Li X, Song W, Liu Y, Huang L, Zhang J, Shen L, Li X, Song W, Liu Y, Huang L (2019) Nanoformulated codelivery of quercetin and alantolactone promotes an antitumor response through synergistic immunogenic cell death for microsatellite-stable colorectal cancer. *ACS Nano* 13(11):12511–12524
64. Helmy W, Sallam M, Fang J, Elkhodairy KA, Elzoghby AO (2018) Dual-targeted casein micelles as green nanomedicine for synergistic phytotherapy of hepatocellular carcinoma. J Control Release [Internet] 287:78–93. https://doi.org/10.1016/j.jconrel.2018.08.026
65. Desai PP, Patravale VB (2017) Curcumin Cocrystal micelles d multifunctional nanocomposites for management of neurodegenerative ailments. J Pharm Sci [Internet] 107:1–14. https://doi.org/10.1016/j.xphs.2017.11.014
66. Xi J, Guo R (2007) Acid – base equilibrium of puerarin in CTAB micelles. J Pharm Biomed Anal 43(April 2006):111–118
67. Zhang L, Chen Z, Yang K, Liu C, Gao J, Qian F (2015) β - lapachone and paclitaxel combination micelles with improved drug encapsulation and therapeutic synergy as novel nanotherapeutics for nqo1-targeted cancer therapy. Mol Pharm. 12(11):3999–4010
68. M AE, A MM (2017) Accepted us [Internet]. Pharmaceutical development and technology, Vol. 0. Taylor & Francis. https://doi.org/10.1080/10837450.2017.1344994
69. Ghaffari M, Dehghan G, Baradaran B, Zarebkohan A, Mansoori B, Soleymani J, et al. Co-delivery of curcumin and Bcl-2 siRNA by PAMAM dendrimers for enhancement of the therapeutic efficacy in HeLa cancer cells. Colloids Surf B Biointerfaces [Internet]: 2019https://doi.org/10.1016/j.colsurfb.2019.110762, 188, 110762
70. Vishnu S, Rompicharla K, Kumari P, Bhatt H, Ghosh B, Biswas S (2018) Biotin functionalized PEGylated poly (amidoamine) dendrimer conjugate for active targeting of paclitaxel in cancer. Int J Pharm [Internet] 557:329–341. https://doi.org/10.1016/j.ijpharm.2018.12.069
71. Shakeri M, Hadi S, Shakeri S (2019) Carvacrol and astaxanthin co-entrapment in beeswax solid lipid nanoparticles as an efficient nano-system with dual antioxidant and anti- bio fi lm activities. LWT - Food Sci Technol [Internet] 107(December 2018):280–290. https://doi.org/10.1016/j.lwt.2019.03.031
72. Ban C, Jo M, Hyun Y, Hwan J, Yong J (2019) Enhancing the oral bioavailability of curcumin using solid lipid nanoparticles. Food Chem [Internet] 302:125328. https://doi.org/10.1016/j.foodchem.2019.125328
73. Press D (2018) Pomegranate extract-loaded solid lipid nanoparticles: design, optimization, and in vitro cytotoxicity study. Int J Nanomedicine 13:1313–1326
74. Singh A, Ahmad I, Akhter S, Jain GK, Iqbal Z, Talegaonkar S et al (2013) Biointerfaces nanocarrier based formulation of thymoquinone improves oral delivery: stability assessment, in vitro and in vivo studies. Colloids SurfB Biointerfaces [Internet] 102:822–832. https://doi.org/10.1016/j.colsurfb.2012.08.038
75. Wang S, Chen T, Chen R, Hu Y, Chen M, Wang Y (2012) Emodin loaded solid lipid nanoparticles : Preparation, characterization and antitumor activity studies. Int J Pharm [Internet]. 430(1–2):238–246. https://doi.org/10.1016/j.ijpharm.2012.03.027
76. Wang X (2017) Capsaicin-loaded nanolipoidal carriers for topical application: design, characterization, and in vitro / in vivo evaluation. Int J Nanomedicine 12:3881–3898
77. Ng WK, Yazan LS, Yap LH, Abd W, Wan G, Hafiza N et al (2015) Thymoquinone-Loaded Nanostructured Lipid Carrier Exhibited Cytotoxicity towards Breast Cancer Cell Lines (MDA-MB-231 and MCF-7) and Cervical Cancer Cell Lines (HeLa and SiHa). Biomed Res Int 2015:263131
78. Fan X, Chen J, Shen Q (2013) Docetaxel – nicotinamide complex-loaded nanostructured lipid carriers for transdermal delivery. Int J Pharm [Internet]. 458(2):296–304. https://doi.org/10.1016/j.ijpharm.2013.10.036
79. Press D (2013) Formulation design, preparation, and in vitro and in vivo characterizations of β -elemene- loaded nanostructured lipid carriers. Int J Nanomedicine 14:2533–2541

80. Chen-yu G, Chun-fen Y, Qi-lu L, Qi T, Yan-wei X, Wei-na L et al (2012) Development of a Quercetin-loaded nanostructured lipid carrier formulation for topical delivery. Int J Pharm 430:292–298
81. Zhao L, Bi D, Qi X, Guo Y, Yue F (2019) Polydopamine-based surface modification of Polydopamine-based surface modification of paclitaxel nanoparticles for osteosarcoma targeted therapy. Nanotechnology 30(25):255101
82. Shanmugam R, Priyanka DL, Madhuri K, Gowthamarajan K, Venkata V, Reddy S et al (2017) Formulation and Characterization of Chitosan Encapsulated Phytoconstituents of Curcumin and Rutin Nanoparticles. Int J Biol Macromol [Internet]. https://doi.org/10.1016/j.ijbiomac.2017.06.112
83. Sahu S, Saraf S (2013) Biocompatible nanoparticles for sustained topical delivery of anticancer phytoconstituent quercetin. Pak J Biol Sci 16:601–609
84. Hsu W, Ng L, Wu T, Lin L, Yen F, Lin C (2012) Characteristics and antioxidant activities of silymarin nanoparticles. J Nanosci Nanotechnol 12(3):2022–2027
85. Pawar AP, Chellampillai B (2011) Andrographolide, a novel bioactive phytoconstituent encapsulated in sustained release biodegradable nanoparticles. Int J Nanotechnol 8:764–778
86. Khurana RK, Sharma T, Kaur S, Jain A, Singh B (2018) 7 role of nanoconstructs in ameliorating chemopreventive and chemotherapeutic potential of plant bioactives. NanoAgroceuticals & NanoPhytoChemicals, 151
87. Bioavailability IV, Monika P, Basavaraj BV, Murthy KNC, Ahalya N, Gurudev K (2017) Nanocapsules of catechin rich extract for enhanced antioxidant potential and nanocapsules of catechin rich extract for enhanced antioxidant potential and in vitro bioavailability. J Appl Pharm Sci *7*(01):184–188
88. Yao J, Zhang Y, Hu Q, Zeng D, Hua F, Meng W et al (2017) PT [Internet]. Eur J Pharm Sci. 101:189. https://doi.org/10.1016/j.ejps.2017.01.028. Elsevier B.V
89. Stecanella LA, Taveira SF, Marreto RN, Valadares MC, De M (2013) Development and characterization of PLGA nanocapsules of grandisin isolated from Virola surinamensis : in vitro release and cytotoxicity studies. Rev Bras Farmacogn – Braz J Pharmacogn [Internet] 23(1):153–159. https://doi.org/10.1590/S0102-695X2012005000128
90. Chen Y, Lin X, Park H, Greever R (2009) Study of artemisinin nanocapsules as anticancer drug delivery systems. Nanomed Nanotechnol Biol Med [Internet] 5(3):316–322. https://doi.org/10.1016/j.nano.2008.12.005
91. Strategies NB (2019) Nanoemulsion and nanoliposome based strategies. *Nutrients 11*(5):1052
92. Mostafa NM (2018) Antibacterial activity of Ginger (Zingiber officinale) leaves essential oil nanoemulsion against the cariogenic streptococcus mutans. J Appl Pharm Sci 8(09):34–41
93. Kaur A, Gupta S, Tyagi A, Sharma RK, Ali J, Gabrani R et al (2017) Development of Nanoemulsion based gel loaded with Phytoconstituents for the treatment of urinary tract infection and in vivo biodistribution studies. Tabriz Univ Med Sci [Internet] 7(4):611–619.. Available from:. https://doi.org/10.15171/apb.2017.073
94. Thomas J, Thanigaivel S, Vijayakumar S, Acharya K, Shinge D, Seelan TSJ et al (2014) Colloids and surfaces B : biointerfaces Pathogenicity of Pseudomonas aeruginosa in Oreochromis mossambicus and treatment using lime oil nanoemulsion. Colloids Surf B Biointerfaces [Internet] 116:372–377. https://doi.org/10.1016/j.colsurfb.2014.01.019
95. Santiago RR, Gyselle K, Holanda D, Dantas N, Genre J, Freitas V et al (2018) Nanostructured lipid carriers containing amphotericin B: development, in vitro release assay, and storage stability. J Drug Deliv Sci Technol [Internet]. https://doi.org/10.1016/j.jddst.2018.10.003
96. Cells ML (2018) Effects of apigenin and apigenin-loaded nanogel on induction of apoptosis in human chronic myeloid leukemia cells. Galen Med J *7*:1008
97. Ashra B, Rashidipour M, Marzban A, Soroush S (2019) Mentha piperita essential oils loaded in a chitosan nanogel with inhibitory e ff ect on bio fi lm formation against S. mutans on the dental surface. Carbohydrate Polym 212(February):142–149
98. Divya G, Panonnummal R, Gupta S, Jayakumar R, Sabitha M (2016) Acitretin and Aloe-emodin loaded chitin nanogel for the treatment of psoriasis Amrita School of Pharmacy, Amrita Institute of Medical Sciences and Research Amrita Centre for Nanosciences and

Molecular Medicine, Amrita Institute of Medical. Eur J Pharm Biopharm [Internet]. https://doi.org/10.1016/j.ejpb.2016.06.019
99. Zhaveh S, Mohsenifar A, Beiki M, Tahereh S, Abdollahi A, Rahmani-cherati T et al (2015) Encapsulation of Cuminum cyminum essential oils in chitosan-caffeic acid nanogel with enhanced antimicrobial activity against Aspergillus flavus. Ind Crop Prod [Internet] 69:251–256. https://doi.org/10.1016/j.indcrop.2015.02.028
100. Tahereh S, Mohsenifar A, Beyki M, Zhaveh S, Rahmani-cherati T, Abdollahi A et al (2015) LWT - food science and technology encapsulation of thyme essential oils in chitosan-benzoic acid nanogel with enhanced antimicrobial activity against Aspergillus flavus. LWT - Food Sci Technol [Internet] 60(1):502–508. https://doi.org/10.1016/j.lwt.2014.07.054
101. Saraf S, Gupta A, Alexander A, Khan J (2015) Advancements and avenues in Nanophytomedicines for better pharmacological responses. J Nanosci Nanotechnol 15:4070
102. Shukla R, Chauhan N, Rajak C, Flora SJS (2019) Flavors and fragrances. In: Flavor development for functional foods and nutraceuticals. CRC Press, Boca Raton
103. Pandey M, Debnath M, Gupta S, Chikara SK (2011) Phytomedicine: an ancient approach turning into future potential source of therapeutics. J Pharmacogn Phyther 3(2):27–37
104. Bento P, Maria K, Negri S (2014) Nanotechnology-based drug delivery systems and herbal medicines: a review. Int J Nanomedicine 9:1–15

Nanophytomedicine: An Effective Way for Improving Drug Delivery and Bioavailability of Herbal Medicines

Mohammad Zaidur Rahman Sabuj and Nazrul Islam

Abstract

Herbal drugs are recognized as a precious gift of nature for the treatment of different diseases and still represent a significant source of many modern medicines. Being eco-friendly, cost-effective formulations, and having considerably low side effects, the herbal medicines have always drawn attention from ancient times to date. However, herbal medicines have high molecular weight with poor solubility that results in poor bioavailability. The nanoparticles (NPs) of drugs have great potential to improve drug solubility and bioavailability. Nanoparticles of plant medicines known as nanophytomedicines have a potential future for enhancing their effectiveness by overcoming problems associated with other forms of plant medicines. Although the bioavailability studies of plant medicines are still in infancy, this chapter provides the updated status of the bioavailability of nanophytomedicines. Additionally, the inhaled route of plant-based nanomedicines delivery has been reported. More studies are warranted to understand the bioavailability of nanophytomedicines to implement the application of plant-based nanomedicines.

Keywords

Nanophytomedicines · Herbal drugs · Nanotechnology · Solubility · Drug delivery · Bioavailability · Anticancer drugs · Lung drug delivery

M. Z. R. Sabuj · N. Islam (✉)
Faculty of Health, Pharmacy Discipline, School of Clinical Sciences, Queensland University of Technology, Brisbane, QLD, Australia
e-mail: mohammadzaidurrahman.sabuj@hdr.qut.edu.au; nazrul.islam@qut.edu.au

S. Beg et al. (eds.), *Nanophytomedicine*,
https://doi.org/10.1007/978-981-15-4909-0_4

1 Introduction

1.1 Nanotechnology and Drug Delivery

The nanoparticles of drugs have a unique characteristic, which lead to enhance their performances in various dosages forms compared to those of large particles. The nanoparticles with large surface area have better access to the solvent molecules around the particles, resulting in faster dissolution and better bioavailability of drug particles [1]. Therefore, the nanoparticles of drugs overcome the solubility problems as the poor water solubility limits the bioavailability of some drugs and affects the performance of the drug formulation. The nanotechnology has proven that the prepared nanoparticles have great potential as drug carriers owing to unique physicochemical and biological properties. The encapsulation of phytomedicines in nanocarriers showed increased solubility and bioavailability of many plant medicines with reduced toxicity [2].

The nanomedicines influence in delivering drugs into the central nervous system. The nanoparticles of drug interact properly with the endothelial cells located in blood–brain barrier (BBB) and produce higher concentration of the dissolved drug in brain cells [3]. The hydrophilic drugs are unable to cross the BBB. Nanomedicine designed with appropriate polymers also helps target drug delivery to a specific organ. Delivery of nanomedicines into brain tumour [4], Alzheimer's disease [5], and CNS disorders [6] has been reported. Additionally, the polymer nanomedicines are considered to be the best candidate for developing the controlled release of various drugs [1, 7, 8].

1.2 Preparation of Nanoparticles/Nanomedicines

Solvent Evaporation Technique The most frequently used two steps technique where initially the emulsification of polymer solution into an aqueous phase is done. The formation of nanoparticles occurred with subsequent evaporation of solvent. The formed nanoparticles are collected either by the filtration or centrifugation technique and dry them using the freeze-drying. The nanoparticles of most polymers, i.e., PLGA, PCL are produced using this technique.

Single Emulsion and Evaporation Method Single emulsion is one of the popular methods for synthesizing polymers micro- or nanoparticles. In this technique, a drug molecule is dissolved in the aqueous phase followed by emulsifying in a non-aqueous medium (i.e., oil). Using a suitable cross-linker, cross-linking of the dispersed droplet is carried out to form a particle. The cross-linking can be achieved both by employing heat and by using the chemical cross-linkers such as glutaraldehyde, formaldehyde or diacid chloride, etc. [9].

Double Emulsion and Evaporation Method An aqueous drug solution is added into the polymer solution in an organic solvent under vigorous stirring to form a w/o emulsion, which is added to a second aqueous phase (surfactant solution) to form a w/o/w emulsion. The encapsulation of protein nanoparticles is produced using this technique.

Solvent Diffusion Technique Using a partially water-miscible solvent like benzyl alcohol, a polymer solution is prepared and saturated this solution with water to establish the thermodynamic stability between both liquids. This polymer solution (now act as a solvent phase) is emulsified in an aqueous solution containing a stabilizer/surfactant. Finally, the solvent diffusion to the external phase is occurred and nanoparticles are formed by evaporating the solvent. Doxorubicin and plasmid DNA loaded PLGA nanoparticles are produced using this technique.

Solvent Displacement/Precipitation Method The solvent displacement involves the precipitation of a preformed polymer from an organic solution and the diffusion of the organic solvent in the aqueous medium with or without the presence of a surfactant. Polymer, drugs, and/ or lipophilic surfactants are dissolved in a semipolar water-miscible solvent (acetone/ethanol), which is added to an aqueous surfactant solution under vigorous magnetic stirring. The nanoparticles of drug are formed spontaneously by rapid diffusion of solvent. The nanoparticles are separated by centrifugation and freeze-dried.

Salting-Out Method This is a modified version of the emulsion process called salting-out method. This method was developed to avoid surfactants and chlorinated solvents, thus avoiding organic solvents to protect the environment and physiological systems [10].

1.3 Nanoparticles of Drugs

Nanoparticles of various drugs are prepared using a variety of natural and synthetic polymers. Nanoparticles of polymer-drug conjugate, polymer encapsulated drug, solid-lipid nanoparticles, and drug-loaded liposomes are very commonly used nanomedicines.

Polymer Nanoparticles The chitosan is the most abundantly available biocompatible and biodegradable polymer and the nanoparticles (NPs) of this polymer as carriers represent promising nanomedicines in modern drug delivery systems ([11, 12]). Using this polymer the development of nicotine and diltiazem loaded chitosan nanocarriers have been successfully investigated for controlled release delivery with better solubility [7, 13]). The chitosan-coated opioid nanoparticles delivered to neuronal cells found to release the drug from NPs up to 48 h [14]. The chitosan NP loaded with opioid protected the drug from degradation and prolonged its intracellular effects. The nanoparticles of other polymers, i.e., polylactic-co-glycolic acid (PLGA), polyethylene glycol (PEG), polylactic acid (PLA), alginate and poly butyl cyanoacrylate (PBCA) for delivering methotrexate, dalargin, temozolomide, lamivudine, zidovudine, and vasoactive intestinal peptide have been reported [15].

Solid-Lipid Nanoparticles (SLNS) The solid-lipid nanoparticles of many drugs are constituted by a matrix of lipids (fatty acids or triglycerides) that remains solid at both room and body temperatures. Here the liquid lipid is replaced by solid-lipid and it improves the solubility of poorly water-soluble drugs resulting in increasing the bioavailability. The lipophilic SLN can cross the BBB [16] without the modification of particle surface and can facilitate the controlled release of drugs [17]. The delivery of SLNs containing anticancer drugs doxorubicin, camptothecin, paclitaxel, melatonin, and etoposide in brain cells has been investigated with significant success [15].

Liposomal Nanoparticles Liposomes are small vesicles of unilamellar or multilamellar phospholipid bilayers. They are most frequently used lipid-based nanocarriers for drug delivery. Owing to the presence of lipid bilayer, liposome can encapsulate both hydrophilic and hydrophobic drugs. Liposomal doxorubicin and daunorubicin produced long half-life with improved toxicity profiles ([18, 19]). Furthermore, the PEGylated liposomal doxorubicin, the first anticancer nanomedicine showed to produce the longest circulating nanomedicine [20]. Considering these, it is proposed that the nanotechnology is an essential technique, which is applicable to the development of nanophytomedicines to combat various diseases in future.

2 Nanophytomedicines and Their Bioavailability Studies

The herbal drugs extracted from various plants are recognized as useful therapeutic agents all over the world since ancient times for various diseases because of their limited adverse effects compared to those of the modern medicines. The phytopharmacological sciences have already showed the efficient biological activities of many plant products [21]. Most of the biologically active plant extracts, i.e., flavonoids, tannins, and terpenoids, are highly water-soluble, however demonstrated a low absorption. Additionally, they have high molecular sizes, which caused to limit their absorption and bioavailability. Owing to the high molecular weight and poor solubility, the majority of plant medicines are incapable of passing the lipid membranes of the cells that leads to their poor bioavailability, which caused problems in clinical trials of the plant products [22]. The recent advances in the herbal revolution have directed the development and delivery of poorly water-soluble herbal medicines for enhancing their bioavailability and efficacy through nanotechnology. The bioavailability studies of some plant medicines, i.e., Apigenin, curcumin, and taxifolin have been reviewed [21]. Although the bioavailability of drug depends on the routes of administration, the nanomedicines offer tremendous advantages of the increased solubility and bioavailability of many drugs. Additionally, the nanophytomedicines help obtain better bioavailability with significantly reduced adverse effects. Therefore, this approach encouraged to include

Table 1 Nanophytomedicines and their bioavailability studies

Plant name	Plant extract	Nanoformulations	Outcomes	References
Curcuma longa	Whole plant	Curcuminoids solid-lipid nanoparticles	Improved bioavailability in brain tissue	Mishra and Palanivelu [26]
Curcuma longa	Whole plant extract	Nanogels	Enhanced solubility and enhanced bioavailability	Gonçalves et al. [27]
Tripterygium wilfordii Hook F	Whole plant extract	Celastrol nanoparticles	Enhanced hydrophilicity	Chen et al. [28]
Boswellia caraterii and *Commiphora myrrha*	Seed extract	Solid-lipid nanoparticles	Improved hydrophobicity and bioavailability	Shi et al. [29]
Scutellaria baicalensis Georgi	Leaf extract	Pulmonary nanocrystal	Enhanced bioavailability	Han et al. [30]
Salvia miltiorrhiza Bunge	Root extract	Solid-lipid nanoparticles	Enhanced bioavailability	Su et al. [31]
Artemisia annua L.	Whole plant extract	Artemisinin nanocapsules	Enhanced bioavailability	Yadav et al. [32]
Panax notoginseng	Root extract	drug-loaded core-shell hybrid liposomal vesicles	Increased protective effect and enhanced bioavailability	Zhang et al. [33]
Taxus brevifolia	Bark extract	Taxol loaded nanoparticles	Enhanced bioavailability	Ma and Mumper [34]
Oroxylum indicum	Flowers extract	Pulmonary nanocrystal	Enhanced bioavailability	Li et al. [35]
Tetradium spp.	Whole plant extract	Evodiamine encapsulated nanoparticles	Enhanced bioavailability	Li et al. [36]
Bergenia spp.	Root extract	Bergenin loaded nanoparticles	Enhanced bioavailability	Rao et al. [37]
Sophora flavescens	Whole plant extract	Oxymatrine loaded nanoparticles	Enhanced bioavailability	Yan et al. [38]

the nanophytomedicines into the novel drug delivery system for various biomedical applications [23]. The lipid nanoemulsion of *Nigella sativa* oil showed to enhance drug solvability and bioavailability [24]. The application of nanophytomedicines in cancer therapy has been extensively reviewed [25]. Here in this section, the bioavailability studies of some nanophytomedicines are presented in Table 1.

2.1 Various Nanophytomedicines

Camptothecin, a quinoline alkaloid is commercially obtained from stem tissues of *Camptotheca acuminate* showed promising anticancer activity. Using a rat lung cancer model, the pulmonary delivery of microparticle of this drug produced significant anticancer efficacy. An orthotopic animal (nude rats) model of lung cancer used in this study and its ability was found to induce lung cancer around 100% [39]. In comparison with the free drug, 9-nitrocamptothecin polymeric nanoparticles showed fivefold increased plasma concentration when tested in a rat model during cytotoxicity [40].

Silymarin, a seed extract, is rich in flavonoid compounds known as flavonolignans [41]. While using in a rat model, it was found to show antioxidant, scavenger, and regulatory effect on the intracellular content of glutathione. It also acted as the membrane stabilizer and permeability regulator to prevent hepatotoxic agents from entering the hepatocytes. In the same experiment it promoted the ribosomal RNA synthesis [42]. However, its water solubility is very poor and an acidic medium is essential for dissolving this drug. The drug solubility was enhanced about threefold in silymarin-loaded polyvinylpyrrolidone nanoparticles comparing to commercially available silymarin product (Legalon®) [43]. Using a rat model, they studied the hepatoprotective effect; however, no bioavailability studies were carried out. It was encapsulated in the Eudragit S 100 (ES100) and Eudragit RL (ERL) polymers to increase the bioavailability.

Danshen is a popular plant medicine extracted from *Radix salvia miltiorrhiza* and commonly used in China. Activities of this nanomedicine are found in coronary heart diseases, myocardial infarction, and angina pectoris. Using a high-speed centrifugal sheering pulverizer, the nanoformulation of this herbal product showed to improve aqueous solubility and bioavailability. It has also improved the quality of drug release profile [31]. Bioavailability of this plant was studied in healthy volunteers and was found that micronization of this plant medicine increased fivefold plasma concentration in comparison with the normal formulations [44].

The extracts from *Phyllanthus amarus Schum and Thonn ameliorates are* traditionally used herbal medicines for hepatoprotectivity. The limitation of the current form of this medicine is that the amount of the herbal remedies required from this plant extract is large, and continuation of the treatment is time consuming. However, ethanolic extract of this plant was encapsulated in sodium alginate for sustained release of constituents in intestine and facilitate maximum absorption. Using a rat model, the efficacy of the nanoencapsulated ethanolic extract of *P. amarus* (NPA) and *P. amarus* (PA) was compared for the hepatoprotective activity. An oral dose of NPA showed a better hepatoprotective activity than PA with reduced toxicity [45].

Cuscuta has been derived from the plant *Cuscuta chinensis*. Upon oral administration, this medicine has poor aqueous solubility and poor absorption rate. However, nanoformulation of this medicine has improved the rate of water solubility. This nanomedicine has hepatoprotective and antioxidants effect in biological systems. Biological activity was tested in rats for hepatotoxicity and improved water solubility was found while the hepatoprotective effect was significant compared to that of ethanolic extract of this plant; however, no bioavailability studies were reported [46].

2.2 Nanophytomedicines with Antibacterial Effect

A large number of medicinal plants/extracts have been recognized for their antibacterial properties. Nowadays the conventional antibiotics are losing their appeal as microbes are growing biofilms around their cell wall against currently available antibiotics. In addition, conventional antibiotics are unable to penetrate the biofilm layer of the antibiotic resistant bacteria. Nanophytomedicine can overcome the limitations with the nanoformulations of various antimicrobial drugs as they can penetrate the biofilm layers.

Artemisinin, a plant medicine, has been extracted from *Artemisia annua* L. Artemisinin and its derivatives can control aggressive cancers, bacterial infection including other parasitic diseases. The isolated artemisinin compounds are known to control the immune response by regulating cell proliferation and cytokine release. However, low bioavailability, short half-life, and limited tissue access were the main limitations with this medicine before making the formulation in nanostructured. Nanoformulations of this plant medicine and its derivatives are still in the clinical trials [47] and a long way to go to be included in the modern formulations.

Berberine is an isoquinoline alkaloid which is extracted from *Berberis aristata* and its several parts are being used in Ayurveda medicine. This plant is traditionally being used for antibacterial, antiperiodic, antidiarrheal, and anticancer activities. Most importantly an anticancer and hepatoprotective agent is found in this plant. It showed low oral bioavailability due to poor aqueous solubility [48]. However, the nanoformulation of this medicine has increased the solubility and dissolution due to the conversion of the crystalline structure to a semi-crystalline form. The nanoparticles of berberine showed better antibacterial and antifungal activities than that of normal berberine formulations. In another study, it was loaded with a solid polymeric particle and found increased dissolution and bioavailability [49].

2.3 Phytosomes

The plant extracts complexed with phosphatidylcholine that forms a new drug delivery system known as phytosome, which showed better phytopharmacological profiles of many plant medicines. The phytosome showed promising bioavailability of various phytomedicine present in milk thistle, grape seed, green tea, olive, and turmeric. This outcome is indicative of the advancement in phytonanomedicine technology and its application in the management of various diseases [50]. Ex-vivo permeation studies of phosphatidylcholinated phytosome system based on the triterpenoid fraction of *Terminalia Arjuna* for efficient delivery of triterpenoid to enhance bioavailability have been demonstrated [51]. The authors explained that the significant absorption of the nanoparticles was due to the lipid miscibility of the compound and suggested that the phytosomes might be considered as an efficient carrier for triterpenoids absorption across the biological membrane and thus can be used for effective management of various diseases.

Celastrol, a promising anticancer herbal drug extracted from the roots of *Celastrus regelii*, is poorly water-soluble and has limited absorption behaviour. The self-assembled phytosomal nanocarriers (178 ± 7 nm) with phospholipid showed significantly improved oral bioavailability (rabbit model) compared to that of the crude drug as evidenced by a fivefold increase in plasma concentration [52]. This result is an indication of the potential application of phytosomal celastrol for oral cancer therapy with improved bioavailability.

2.4 Nanophytomedicines in Cancer Therapy

The use of the nanoparticulate drug has been found to resolve various challenges in drug delivery especially to the cancer cells. This includes preferential accumulation of the nanoparticles into the cancer cells, extending half-lives of drugs in the circulation, and reducing unwanted adverse effects in non-target organs. Nanoparticulate delivery of many anticancer drugs has been reported with a significant success [53, 54]. Nature has provided us with potent anticancer drugs such as taxanes (paclitaxel, docetaxel), vinca alkaloids, podophyllotoxin and its derivatives, camptothecin and its derivatives.

Paclitaxel, the most promising herbal anticancer drug, extracted from the bark of the Pacific yew (*Taxus brevifolia* L.). Initially, this drug was approved for clinical use against ovarian and breast cancers. The paclitaxel in albumin nanoparticle formulation (Abraxane) showed to increase the solubility and the bioavailability resulting in a higher intratumour concentration of the drug [55]. The PEGylated paclitaxel nanoparticles were studies for lung cancer [53]. The chitosan-based paclitaxel NPs were found to be preferentially accumulated in lung cancer A549 cells and showed a significant inhibition in cell proliferation [56]. Using a rat lung model, the chitosan modified PLGA NPs loaded with paclitaxel showed significantly increased uptake and cytotoxicity against A549 lung cancer cells [57]; however, no bioavailability studies were carried out.

Vincristine, a dimeric Catharanthus alkaloid was derived from the Madagascar periwinkle [58]. It has effects on uncontrolled cell division by binding to tubulin protein and blocking the metaphase stage of cell division. These characteristics of vincristine were used to develop treatments for several human cancers. If the liposomal system is developed properly, then vincristine can be encapsulated in them where significant therapeutic effects are found. Drug toxicity can also be slightly decreased using enhanced efficacy of vincristine [59]. Using a rat model, the oral bioavailability of vincristine from Dextran-PLGA hybrid nanoparticles was increased up to 3.3-fold compared to that of the drug solution [60]. The encapsulation of the drug in the nanoparticles was carried out by forming electrostatic complex.

Vinblastine is the first therapeutic agent from plants with high anticancer properties. Vinblastine loaded poly(ethylene glycol)-folate was showed to release the polymeric nanoparticles within 2 days under a mimic tumour intracellular condition. However, in vivo drug accumulation in tumor cell was high in concentration but low release in normal tissues or blood, thus decreased toxicity in normal tissues [61].

Docetaxel is another routinely used plant based anticancer drug against various types of cancers. This drug produces its cytotoxic effect by interfering with the tubulin formation. This drug (Taxotere®) is administered intravenously, and known to cause serious adverse effects such as cardiotoxicity and nephrotoxicity. The docetaxel-loaded chitosan microspheres showed maximum drug accumulation in a mouse model lung cancer cells, with a minimal exposure into the healthy tissues; however, the mechanism of drug accumulation in to the cancer cells was not demonstrated [62]. This drug encapsulated in chitosan microspheres produced a sustained drug release up to 19 days.

Curcumin is a very well-known polyphenol derived from the plant *Curcuma longa* and has a long history of its application as herbal medicine for managing various diseases. The bioavailability of curcumin from the currently available dosage forms is very poor owing to its poor water solubility and poor absorption. The curcumin loaded PEGylated SLNs showed rapid permeability through the epithelium, resulting in a 12-fold increase in bioavailability compared to that of curcumin solution in a rat model [63]. In another study, the oral delivery of curcumin loaded SLN with mesoporous silica shell produced enhanced bioavailability of curcumin [64]. In another study, the orally administered curcumin loaded SLNs in a rat model showed 39 times higher plasma concentration of curcumin compared to the curcumin at 50 mg/kg dose [65]. Peng et al. loaded the curcumin into sophorolipid micelles and both in vitro and in vivo studies showed that the curcumin nanoparticles had higher bioavailability (three- to fourfold) than that of free curcumin crystals [66]. These are very promising outcomes of curcumin nanoparticles for better bioavailability. The phytosomal curcumin showed promising activity against breast cancer; however, no bioavailability studies were performed [67]. The curcumin loaded PLGA nanoparticles exhibited about 640-fold higher water solubility compared to that of curcumin crystal. The orally administered particles showed 5.6-fold higher bioavailability compared to that of curcumin [68]. Curcumin micellar nanoparticles (Cur-NPs) showed effective cytotoxicity against lung cancer (A549) [69]. It should be noted here that the Cur-NPs were found to be more effective against lung cancer cells compared to that of curcumin alone. Moreover, no nanoparticluate formulations showed toxicity to normal cells (BEAS-2B). Very recently, chitosan-coated curcumin nanoparticles showed tenfold higher bioavailability than that of without chitosan coating [70].

*Ginseng, a t*housand years old Chinese medicine, derived from *Panax ginseng* has high medicinal values. However, the complex components of this herbal product may interact with each other and thus reduce the therapeutic effects. The nanoformulations of this medicine demonstrated high drug loading efficiency, long half-time in the systemic blood circulation, better antitumor effects, higher accumulation at the cancer cells, and reduced toxicity to normal cells. The bioavailability was increased about ninefold comparing with free ginseng in a mice model [71].

Nanoformulations of plumbagin nanoparticles were prepared from the root extract of *Plumbago zeylanica*. It is a very effective herbal preparation for the treatment of prostate cancer. This crude extract is highly toxic for normal cells while nanoformulation is less toxic, thus confirming the cytocompatibility of nano plumbagin with normal cells and dose-dependent toxicity to cancer cells. In vitro studies revealed that plumbagin nanoparticles are biocompatible with blood cells. Its activity was also found for wound healing, while prostate cancer cells migration was interrupted with their activities [72].

3 Lung Delivery of Nanophytomedicines

Direct delivery of drugs into the lungs enables the administration of lower doses with an equivalent therapeutic benefit compared to that of conventional oral route because of the large surface area (~140 m^2) with extremely thin (0.1–0.2 μm) absorptive pulmonary epithelium and good blood supply of the lungs. Due to the high surface area of nanoparticles that provides sufficient area for drugs to be dissolved and absorbed, the bioavailability of the delivered drugs is increased. The pulmonary drug delivery is a non-invasive delivery route that has superior efficacy, and rapid onset of action, which other drug administration routes (except injectable products) is unable to achieve. Additionally, lung delivery of drugs can avoid fast pass metabolism, which is a common problem for drug formulations administered orally. Moreover, deep lung delivery of drugs ensures low systemic adverse effects compared with those of oral dosage forms. The inhaled delivery of nanoparticulate drug offers enormous benefits in the management of many diseases. Regarding the anticancer drugs, the lung delivery offers an attractive therapeutic route to efficient delivery of drugs into the lung cancer cells. Therefore, the lung delivery of anticancer drugs is likely to increase the treatment efficiency without affecting the normal cells [73].

The biocompatible and biodegradable chitosan treated PLGA NPs loaded with paclitaxel showed significantly increased cell uptake and cytotoxicity against cancer cells in a rat model [57]; however, no bioavailability studies were carried out. The authors suggested that the presence of the positively charged chitosan on the surface of NPs caused to electrostatic interaction with negatively charged tumour cells. This was ample justify the cancer cell specific accumulation of the administered nanoparticles. The PEGylated paclitaxel was reported to be retained within the cancer cells and found to improve its efficacy with the reduction of toxicity compared to that of intravenously administered taxol [74]. Intratracheal instillation of the PEGylated paclitaxel NPs showed to be retained within the cancer cells with controlled release of the drug that resulted in improving its efficacy [53]. Overall, the outcomes from these studies are very encouraging and confirmed the anticancer activity of the inhaled paclitaxel nanoparticles. Using various animal models, more exclusive experiments are warranted to confirm the bioavailability of this drug upon inhalation. The inhaled formulation of liposomal etoposide and docetaxel nanoparticle showed the sustained drug release in the cancer cells significant cytotoxic effect against H-1299 and A-549 cells [75]. The prepared drug-loaded cationic liposomes showed promising intracellular uptake with enhanced cytotoxicity.

Yuanhuacine, a herbal medicine, isolated from the flower buds of *Daphne genkwa*, has been demonstrated to have anticancer activity against various cancer cells. The inhaled powder formulation of this drug has been found direct accumulation of drug into the target cells, with reduced distribution to the other organs [76]. This outcome is very promising and the inhaled nanophytomedicine can be applied for the treatment of lung cancers. Further studies are warranted to explore the unwanted adverse effects of this drug on the normal cells. Chen et al. demonstrated the pulmonary delivery of the drug encapsulated PLGA nanoparticles NPs [77]. The authors prepared the nanoparticles bonded with RGDfk ligand and histidine groups on the surface to facilitate the interaction between the drug and the integrin $\alpha_v\beta_3$ receptor on the cancer cells.

Oridonin, a diterpenoid extracted from the herb *Rabdosia rubescens*, is well-known as a potent anticancer drug. The inhaled oridonin-loaded poly(lactic-co-glycolic) acid (PLGA) particles efficiently deposited into the lower level of lungs and produced very strong anticancer activity [78] due to the preferential accumulation of the delivered drug into the cancer cells; however, no bioavailability studies were reported. The spray dried inhalable curcumin powders showed very good dispersibility and the deposited drug accumulated into lung cells [79]. Furthermore, the spray dried powder formulation showed significantly higher solubility compared to that of the original curcumin. The enhanced bioavailability of NPs upon inhalation in a mouse model was also demonstrated [79].

The lung delivery of anticancer drug nanoparticles showed to accumulate in the cancer cells. The animal models showed promising bioavailability with increased drug concentration in the lungs upon inhalation. Additionally, the drug accumulation into other organs such as liver, kidney, and reproductive systems was greatly reduced, which is an indication of very low toxicity. Therefore, the application of the nanophytomedicines is not only confined to the oral or injectable routes but also extended application via a pulmonary route with high efficiency at a greatly reduced dose.

4 Conclusion

The nanophytomedicine proved to be a very useful form of plant medicines for the management of various diseases with improved bioavailability profile and less toxicity. As the herbal medicines have high molecular weight with poor solubility that results in poor bioavailability, the nanotechnology can contribute to increase the solubility of highly lipophilic drugs resulting in increasing the bioavailability. This is a potential surrogate between green medicines and nanotechnology with realistic prospects for managing very life-threatening diseases such as cancer and minimizing the application of modern medicines. The pulmonary drug delivery is a non-invasive delivery route that has superior efficacy, rapid onset of drug action, and can avoid fast pass metabolism, which is a common problem for oral dosage forms of the drug. Thus, the inhaled nanophytomedicines likely have great potential to produce better bioavailability and therapeutic action at a very low dose of drugs. Although some nanophytomedicines have great potential to contribute to the improved bioavailability, very few nanoparticulate natural compounds are being tested in clinical trials. More extensive studies are warranted to understand the bioavailability and toxicities of nanophytomedicines in animal models.

References

1. Singh R, Lillard JW (2009) Nanoparticle-based targeted drug delivery. Exp Mol Pathol 86 (3):215–223
2. Lekshmi NG, Subash CMP, Sundar WA (2018) Role of nanotechnology in herbal medicine. Indo Am J Pharm Sci 5(11):12053–12063. https://doi.org/10.5281/zenodo.1486099
3. McCarthy DJ, Malhotra M, O'Mahony AM, Cryan JF, O'Driscoll CM (2015) Nanoparticles and the blood-brain barrier: advancing from in-vitro models towards therapeutic significance. Pharm Res 32(4):1161–1185. https://doi.org/10.1007/s11095-014-1545-6
4. Chen Y-C, Chiang C-F, Wu S-K, Chen L-F, Hsieh W-Y, Lin W-L (2015) Targeting microbubbles-carrying TGFβ1 inhibitor combined with ultrasound sonication induce BBB/BTB disruption to enhance nanomedicine treatment for brain tumors. J Control Release 211:53–62. https://doi.org/10.1016/j.jconrel.2015.05.288
5. Andrieux K, Couvreur P (2013) Nanomedicine as a promising approach for the treatment and diagnosis of brain diseases: the example of Alzheimer's disease. Ann Pharm Fr 71(4):225–233. https://doi.org/10.1016/j.pharma.2013.04.001
6. Peluffo H, Unzueta U, Negro-Demontel ML, Xu Z, Vaquez E, Ferrer-Miralles N, Villaverde A (2015) BBB-targeting, protein-based nanomedicines for drug and nucleic acid delivery to the CNS. Biotechnol Adv 33(2):277–287. https://doi.org/10.1016/j.biotechadv.2015.02.004
7. Muhsin MDA, George G, Beagley K, Ferro V, Wang H, Islam N (2016) Effects of chemical conjugation of L-leucine to chitosan on dispersibility and controlled release of drug from a nanoparticulate dry powder inhaler formulation. Mol Pharm 13(5):1455–1466. https://doi.org/10.1021/acs.molpharmaceut.5b00859
8. Wang H, George G, Bartlett S, Gao C, Islam N (2017) Nicotine hydrogen tartrate loaded chitosan nanoparticles: Formulation, characterization and in vitro delivery from dry powder inhaler formulation. Eur J Pharm Biopharm 113:118–131. https://doi.org/10.1016/j.ejpb.2016.12.023
9. Rao JP, Geckeler KE (2011) Polymer nanoparticles: preparation techniques and size-control parameters. Prog Polym Sci 36(7):887–913. https://doi.org/10.1016/j.progpolymsci.2011.01.001
10. Bindschaedler C, Gurny R, Doelker E (1990) Process for preparing a powder of water-insoluble polymer which can be redispersed in a liquid phase, the resulting powder and utilization thereof. US Patent 4968350A
11. Islam N, Ferro V (2016) Recent advances in chitosan-based nanoparticulate pulmonary drug delivery. Nanoscale 8:14341–14358. https://doi.org/10.1039/c6nr03256g
12. Islam N, Wang H, Islam N, Maqbool F, Ferro V, Ferro V (2019) In vitro enzymatic digestibility of glutaraldehyde-crosslinked chitosan nanoparticles in lysozyme solution and their applicability in pulmonary drug delivery. Molecules 24(7):1271. https://doi.org/10.3390/molecules24071271

13. Wang H, George G, Islam N (2018) Nicotine-loaded chitosan nanoparticles for dry powder inhaler (DPI) formulations - Impact of nanoparticle surface charge on powder aerosolization. Adv Powder Technol 29(12):3079–3086. https://doi.org/10.1016/j.apt.2018.08.011
14. Malatesta M, Galimberti V, Cisterna B, Costanzo M, Biggiogera M, Zancanaro C (2014) Chitosan nanoparticles are efficient carriers for delivering biodegradable drugs to neuronal cells. Histochem Cell Biol 141(5):551–558. https://doi.org/10.1007/s00418-013-1175-9
15. Wong HL, Wu XY, Bendayan R (2012) Nanotechnological advances for the delivery of CNS therapeutics. Adv Drug Deliv Rev 64(7):686–700. https://doi.org/10.1016/j.addr.2011.10.007
16. Fenart L, Casanova A, Dehouck B, Duhem C, Slupek S, Cecchelli R, Betbeder D (1999) Evaluation of effect of charge and lipid coating on ability of 60-nm nanoparticles to cross an in vitro model of the blood-brain barrier. J Pharmacol Exp Ther 291(3):1017–1022
17. Mueller RH, Ruehl D, Runge S, Schulze-Forster K, Mehnert W (1997) Cytotoxicity of solid lipid nanoparticles as a function of the lipid matrix and the surfactant. Pharm Res 14 (4):458–462
18. Harrison M, Tomlinson D, Stewart S (1995) Liposomal-entrapped doxorubicin: an active agent in AIDS-related Kaposi's sarcoma. J Clin Oncol 13(4):914–920
19. Okochi H, Nakano M (1998) Enhancement of antitumor activity and reduction of adverse reactions by injection of liposomes entrapping doxorubicin or daunorubicin. Igaku Ayumi 184 (5):320–324
20. Bertrand N, Simard P, Leroux J-C (2010) Serum-stable, long-circulating, pH-sensitive PEGylated liposomes. Methods Mol Biol 605:545–558. https://doi.org/10.1007/978-1-60327-360-2_36
21. Watkins R, Wu L, Zhang C, Davis RM, Xu B (2015) Natural product-based nanomedicine: recent advances and issues. Int J Nanomedicine 10:6055–6074. https://doi.org/10.2147/ijn.s92162
22. Bonifacio Bruna V, Silva Patricia Bento D, Ramos Matheus Aparecido Dos S, Negri Kamila Maria S, Bauab Tais M, Chorilli M (2014) Nanotechnology-based drug delivery systems and herbal medicines: a review. Int J Nanomedicine 9:1–15. https://doi.org/10.2147/IJN.S52634
23. Alexander A, Patel Ravish J, Saraf S, Saraf S (2016) Recent expansion of pharmaceutical nanotechnologies and targeting strategies in the field of phytopharmaceuticals for the delivery of herbal extracts and bioactives. J Control Release 241:110–124. https://doi.org/10.1016/j.jconrel.2016.09.017
24. Tabassum H, Ahmad IZ (2018) Evaluation of the anticancer activity of sprout extract-loaded nanoemulsion of N. sativa against hepatocellular carcinoma. J Microencapsul 35(7-8):643–656. https://doi.org/10.1080/02652048.2019.1571641
25. Vohra K, Dureja H, Garg V, Dutt R (2019) Current trends in phyto-cancer therapy using nanoparticles. Curr Nanomed 9(1):30–45. https://doi.org/10.2174/2468187308666180613102448
26. Mishra S, Palanivelu K (2008) The effect of curcumin (turmeric) on Alzheimer's disease: an overview. Ann Indian Acad Neurol 11(1):13–19. https://doi.org/10.4103/0972-2327.40220
27. Gonçalves C, Schellenberg PM, Gama F, Coutinho PJ, Pereira P (2012) Self-assembled dextrin nanogel as curcumin delivery system. J Biomater Nanobiotechnol 3(2):178–184. https://doi.org/10.4236/jbnb.2012.32024
28. Chen S-R, Dai Y, Zhao J, Lin L, Wang Y, Wang Y (2018) A mechanistic overview of triptolide and celastrol, natural products from tripterygium wilfordii Hook F. Front Pharmacol 9:104–104. https://doi.org/10.3389/fphar.2018.00104
29. Shi F, Zhao J-H, Liu Y, Wang Z, Zhang Y-T, Feng N-P (2012) Preparation and characterization of solid lipid nanoparticles loaded with frankincense and myrrh oil. Int J Nanomedicine 7:2033–2043. https://doi.org/10.2147/IJN.S30085
30. Han L, Fu Y, Cole AJ, Liu J, Wang J (2012) Co-encapsulation and sustained-release of four components in ginkgo terpenes from injectable PELGE nanoparticles. Fitoterapia 83 (4):721–731. https://doi.org/10.1016/j.fitote.2012.02.014

31. Su YL, Fu ZY, Zhang JY, Wang WM, Wang H, Wang YC, Zhang QJ (2008) Microencapsulation of Radix salvia miltiorrhiza nanoparticles by spray-drying. Powder Technol 184 (1):114–121. https://doi.org/10.1016/j.powtec.2007.08.014
32. Yadav D, Suri S, Choudhary A, Sikender M, Hemant BN, Beg NJIJPT (2011) Novel approach: herbal remedies and natural products in pharmaceutical science as nano drug delivery systems. Int J Pharm Technol 3(3):3092–3116
33. Zhang J, Han X, Li X, Luo Y, Zhao H, Yang M, Ni B, Liao Z (2012) Core-shell hybrid liposomal vesicles loaded with panax notoginsenoside: preparation, characterization and protective effects on global cerebral ischemia/reperfusion injury and acute myocardial ischemia in rats. Int J Nanomedicine 7:4299–4310. https://doi.org/10.2147/IJN.S32385
34. Ma P, Mumper RJ (2013) Paclitaxel nano-delivery systems: a comprehensive review. J Nanomed Nanotechnol 4(2):1000164. https://doi.org/10.4172/2157-7439.1000164
35. Li X, Luo W, Ng TW, Leung PC, Zhang C, Leung KC-F, Jin L (2017) Nanoparticle-encapsulated baicalein markedly modulates pro-inflammatory response in gingival epithelial cells. Nanoscale 9(35):12897–12907. https://doi.org/10.1039/c7nr02546g
36. Li C, Cai G, Song D, Gao R, Teng P, Zhou L, Ji Q, Sui H, Cai J, Li Q, Wang Y (2019) Development of EGFR-targeted evodiamine nanoparticles for the treatment of colorectal cancer. Biomater Sci 7(9):3627–3639. https://doi.org/10.1039/c9bm00613c
37. Rao K, Roome T, Aziz S, Razzak A, Abbas G, Imran M, Jabri T, Gul J, Hussain M, Sikandar BJJ (2018) Bergenin loaded gum xanthan stabilized silver nanoparticles suppress synovial inflammation through modulation of the immune response and oxidative stress in adjuvant induced arthritic rats. J Mater Chem B 6(27):4486–4501. https://doi.org/10.1039/C8TB00672E
38. Yan C, Chen D, Gu J, Hu H, Zhao X, Qiao M (2006) Preparation of N-succinyl-chitosan and its physical-chemical properties as a novel excipient. Yakugaku Zasshi 126(9):789–793
39. Chao P, Deshmukh M, Kutscher HL, Gao D, Rajan SS, Hu P, Laskin DL, Stein S, Sinko PJ (2010) Pulmonary targeting microparticulate camptothecin delivery system: anticancer evaluation in a rat orthotopic lung cancer model. Anti-Cancer Drugs 21(1):65–76. https://doi.org/10.1097/CAD.0b013e328332a322
40. Dadashzadeh S, Derakhshandeh K, Shirazi FHJA (2008) 9-nitrocamptothecin polymeric nanoparticles: cytotoxicity and pharmacokinetic studies of lactone and total forms of drug in rats. Anti-Cancer Drugs 19(8):805–811. https://doi.org/10.1097/CAD.0b013e3283099e5c
41. Muriel P, Mourelle M (1990) Prevention by silymarin of membrane alterations in acute CCl4 liver damage. J Appl Toxicol 10(4):275–279
42. Sonnenbichler J, Zetl I (1986) Biochemical effects of the flavonolignane silibinin on RNA, protein and DNA synthesis in rat livers. Prog Clin Biol Res 213:319–331
43. Yang KY, Hwang DH, Yousaf AM, Kim DW, Shin Y-J, Bae O-N, Kim Y-I, Kim JO, Yong CS, Choi H-G (2013) Silymarin-loaded solid nanoparticles provide excellent hepatic protection: physicochemical characterization and in vivo evaluation. Int J Nanomedicine 8:3333–3343. https://doi.org/10.2147/IJN.S50683
44. Xing L, Tan Z-R, Cheng J-L, Huang W-H, Zhang W, Deng W, Yuan C-S, Zhou H-H (2017) Bioavailability and pharmacokinetic comparison of tanshinones between two formulations of Salvia miltiorrhiza in healthy volunteers. Sci Rep 7(1):4709–4709. https://doi.org/10.1038/s41598-017-02747-4
45. Deepa V, Sridhar R, Goparaju A, Reddy PN, Murthy PB (2012) Nanoemulsified ethanolic extract of Pyllanthus amarus Schum & Thonn ameliorates CCl4 induced hepatotoxicity in Wistar rats. Indian J Exp Biol 50(11):785–794
46. Yen F-L, Wu T-H, Lin L-T, Cham T-M, Lin C-C (2008) Nanoparticles formulation of Cuscuta chinensis prevents acetaminophen-induced hepatotoxicity in rats. Food Chem Toxicol 46 (5):1771–1777. https://doi.org/10.1016/j.fct.2008.01.021
47. Charlie-Silva I, Fraceto LF, de Melo NFS (2018) Progress in nano-drug delivery of artemisinin and its derivatives: towards to use in immunomodulatory approaches. Artif Cells Nanomed Biotechnol 46(suppl 3):S611–S620. https://doi.org/10.1080/21691401.2018.1505739

48. Sahibzada MUK, Sadiq A, Faidah HS, Khurram M, Amin MU, Haseeb A, Kakar M (2018) Berberine nanoparticles with enhanced in vitro bioavailability: characterization and antimicrobial activity. Drug Des Devel Ther 12:303–312. https://doi.org/10.2147/DDDT.S156123
49. Jia J, Zhang K, Zhou X, Zhou D, Ge F (2018) Precise dissolution control and bioavailability evaluation for insoluble drug berberine via a polymeric particle prepared using supercritical CO_2. Polymers 10(11):1198. https://doi.org/10.3390/polym10111198
50. Ravi GS, Narayana Charyulu R, Dubey A, Hebbar S, Mathias AC (2018) Phytosomes: a novel molecular nano complex between phytomolecule and phospholipid as a value added herbal drug delivery system. Int J Pharm Sci Rev Res 51(1):84–90
51. Shende MA, More MS, Marathe RP (2018) Development and evaluation of Terminalia Arjuna loaded phytosome for bioavailability enhancement. Int J Pharm Sci Nanotechnol 11(2):1
52. Freag MS, Saleh WM, Abdallah OY (2018) Self-assembled phospholipid-based phytosomal nanocarriers as promising platforms for improving oral bioavailability of the anticancer celastrol. Int J Pharm 535(1-2):18–26. https://doi.org/10.1016/j.ijpharm.2017.10.053
53. Meenach SA, Anderson KW, Zach Hilt J, McGarry RC, Mansour HM (2013) Characterization and aerosol dispersion performance of advanced spray-dried chemotherapeutic PEGylated phospholipid particles for dry powder inhalation delivery in lung cancer. Eur J Pharm Sci 49 (4):699–711. https://doi.org/10.1016/j.ejps.2013.05.012
54. Wicki A, Witzigmann D, Balasubramanian V, Huwyler J (2015) Nanomedicine in cancer therapy: challenges, opportunities, and clinical applications. J Control Release 200:138–157. https://doi.org/10.1016/j.jconrel.2014.12.030
55. Harshita BMA, Beg S, Pottoo FH, Ahmad FJ (2019) Nanopaclitaxel therapy: an evidence based review on the battle for next-generation formulation challenges. Nanomedicine 14 (10):1323–1341
56. Jiang J, Liu Y, Wu C, Qiu Y, Xu X, Lv H, Bai A, Liu X (2017) Development of drug-loaded chitosan hollow nanoparticles for delivery of paclitaxel to human lung cancer A549 cells. Drug Dev Ind Pharm 43(8):1304–1313. https://doi.org/10.1080/03639045.2017.1318895
57. Yang R, Yang S-G, Shim W-S, Cui F, Cheng G, Kim I-W, Kim D-D, Chung S-J, Shim C-K (2009) Lung-specific delivery of paclitaxel by chitosan-modified PLGA nanoparticles via transient formation of microaggregates. J Pharm Sci 98(3):970–984. https://doi.org/10.1002/jps.21487
58. Barkat MA, Abul H, Rahman MA (2017) Agricultural, pharmaceutical, and therapeutic interior of Catharanthus roseus (L.). In: Don G (ed) Catharanthus roseus. Springer, Basel, pp 71–100
59. Waterhouse DN, Madden TD, Cullis PR, Bally MB, Mayer LD, Webb MS (2005) Preparation, characterization, and biological analysis of liposomal formulations of vincristine. Methods Enzymol 391:40–57
60. Ling G, Zhang P, Zhang W, Sun J, Meng X, Qin Y, Deng Y, He Z (2010) Development of novel self-assembled DS-PLGA hybrid nanoparticles for improving oral bioavailability of vincristine sulfate by P-gp inhibition. J Control Release 148(2):241–248. https://doi.org/10.1016/j.jconrel.2010.08.010
61. Zhu Y, Liu R, Huang H, Zhu Q (2019) Vinblastine-loaded nanoparticles with enhanced tumor-targeting efficiency and decreasing toxicity: developed by one-step molecular imprinting process. Mol Pharm 16(6):2675–2689. https://doi.org/10.1021/acs.molpharmaceut.9b00243
62. Wang H, Xu Y, Zhou X (2014) Docetaxel-loaded chitosan microspheres as a lung targeted drug delivery system: in vitro and in vivo evaluation. Int J Mol Sci 15(3):3519–3532. https://doi.org/10.3390/ijms15033519
63. Ban C, Jo M, Park YH, Kim JH, Han JY, Lee KW, Kweon D-H, Choi YJ (2020) Enhancing the oral bioavailability of curcumin using solid lipid nanoparticles. Food Chem 302:125328. https://doi.org/10.1016/j.foodchem.2019.125328
64. Kim S, Diab R, Joubert O, Canilho N, Pasc A (2016) Core-shell microcapsules of solid lipid nanoparticles and mesoporous silica for enhanced oral delivery of curcumin. Colloids Surf B Biointerfaces 140:161–168. https://doi.org/10.1016/j.colsurfb.2015.12.040

65. Kakkar V, Singh S, Singla D, Kaur IP (2011) Exploring solid lipid nanoparticles to enhance the oral bioavailability of curcumin. Mol Nutr Food Res 55(3):495–503. https://doi.org/10.1002/mnfr.201000310
66. Peng C, Zheng J, Chen D, Zhang X, Deng L, Chen Z, Wu L (2018) Response of hPDLSCs on 3D printed PCL/PLGA composite scaffolds in vitro. Mol Med Rep 18(2):1335–1344. https://doi.org/10.3892/mmr.2018.9076
67. Hashemzehi M, Behnam-Rassouli R, Hassanian SM, Moradi-Binabaj M, Moradi-Marjaneh R, Rahmani F, Fiuji H, Jamili M, Mirahmadi M, Boromand N, Piran M, Jafari M, Sahebkar A, Avan A, Khazaei M (2018) Phytosomal-curcumin antagonizes cell growth and migration, induced by thrombin through AMP-Kinase in breast cancer. J Cell Biochem 119 (7):5996–6007. https://doi.org/10.1002/jcb.26796
68. Xie X, Tao Q, Zou Y, Zhang F, Guo M, Wang Y, Wang H, Zhou Q, Yu S (2011) PLGA nanoparticles improve the oral bioavailability of curcumin in rats: characterizations and mechanisms. J Agric Food Chem 59(17):9280–9289. https://doi.org/10.1021/jf202135j
69. Lee W-H, Loo C-Y, Ong H-X, Traini D, Young PM, Rohanizadeh R (2016) Synthesis and characterization of inhalable flavonoid nanoparticle for lung cancer cell targeting. J Biomed Nanotechnol 12(2):371–386. https://doi.org/10.1166/jbn.2016.2162
70. Kar SK, Akthar F, Ray G, Pandey AK (2018) Curcumin nanoparticles with improved bioavailability and methods of producing the same. IN Patent 2008-DE1827
71. Dai L, Zhu W, Si C, Lei J (2018) "Nano-Ginseng" for enhanced cytotoxicity against cancer cells. Int J Mol Sci 19(2):627. https://doi.org/10.3390/ijms19020627
72. Nair HA, Snima KS, Kamath RC, Nair SV, Lakshmanan V-K (2015) Plumbagin nanoparticles induce dose and pH dependent toxicity on prostate cancer cells. Curr Drug Deliv 12(6):709–716
73. Islam N, Richard D (2019) Inhaled micro/nanoparticulate anticancer drug formulations: an emerging targeted drug delivery strategy for lung cancers. Curr Cancer Drug Targets 19 (3):162–178. https://doi.org/10.2174/1568009618666180525083451
74. Luo T, Loira-Pastoriza C, Patil HP, Ucakar B, Muccioli GG, Bosquillon C, Vanbever R (2016) PEGylation of paclitaxel largely improves its safety and anti-tumor efficacy following pulmonary delivery in a mouse model of lung carcinoma. J Control Release 239:62–71. https://doi.org/10.1016/j.jconrel.2016.08.008
75. Jinturkar KA, Anish C, Kumar MK, Bagchi T, Panda AK, Misra AR (2012) Liposomal formulations of etoposide and docetaxel for p53 mediated enhanced cytotoxicity in lung cancer cell lines. Biomaterials 33(8):2492–2507. https://doi.org/10.1016/j.biomaterials.2011.11.067
76. Chen Z, Li M, Fan Q, Li W, Duan J (2017) Yuanhuacine dry powder inhaler having anti-lung cancer activity. CN Patent 2016-11016465
77. Chen R, Xu L, Fan Q, Li M, Wang J, Wu L, Li W, Duan J, Chen Z (2017) Hierarchical pulmonary target nanoparticles via inhaled administration for anticancer drug delivery. Drug Deliv 24(1):1191–1203. https://doi.org/10.1080/10717544.2017.1365395
78. Zhu L, Li M, Liu X, Du L, Jin Y (2017) Inhalable oridonin-loaded poly(lactic-co-glycolic)acid large porous microparticles for in situ treatment of primary non-small cell lung cancer. Acta Pharm Sin B 7(1):80–90
79. Hu L, Kong D, Hu Q, Gao N, Pang S (2015) Evaluation of high-performance curcumin nanocrystals for pulmonary drug delivery both in vitro and in vivo. Nanoscale Res Lett 10 (1):1–9. https://doi.org/10.1186/s11671-015-1085-y

Self-Nanoemulsifying Drug Delivery System for Improving Efficacy of Bioactive Phytochemicals

Javed Ahmad, Saima Amin, Sanjeev Singh, Gulam Mustafa, and Md Abul Barkat

Abstract

Self-nanoemulsifying drug delivery systems (SNEDDS) is isotropic mixture of optimum ratio of oils, surfactants, and co-surfactants, which emulsify to nanostructure conditions under gentle agitation encountered in gastro-intestinal tract carrying poorly soluble therapeutics. Most of the therapeutics of herbal origin (like ginkgolide, oleanolic acid, silymarin and curcumin, etc.) are of hydrophobic, so SNEDDS might be the promising carrier for the effective oral delivery to improve bioavailability and reduction in pharmacokinetic variation. When such formulation is released into gut lumen, it disperses to form nanoemulsion, so that therapeutics remains in state of fine dispersion in the GIT, avoiding the dissolution step which frequently limits the rate of absorption of herbal origin drug of hydrophobic nature. This book chapter is aimed to discuss the overview of the formulation design of SNEDDS and applicability of SNEDDS for enhancement of the oral bioavailability/efficacy of bioactive phytochemicals.

J. Ahmad (✉)
Department of Pharmaceutics, College of Pharmacy, Najran University, Najran, Kingdom of Saudi Arabia

S. Amin
Department of Pharmaceutics, School of Pharmaceutical Education and Research, Jamia Hamdard, New Delhi, India

S. Singh
National Institute of Pharmaceutical Education and Research (NIPER), Hajipur, Bihar, India

G. Mustafa
College of Pharmacy (Boys), Al-Dawadmi Campus, Shaqra University, Riyadh, Kingdom of Saudi Arabia

M. A. Barkat
Department of Pharmaceutics, College of Pharmacy, University of Hafr Al-Batin, Hafr Al-Batin, Saudi Arabia

S. Beg et al. (eds.), *Nanophytomedicine*,
https://doi.org/10.1007/978-981-15-4909-0_5

Keywords

SNEDDS · Surfactants · Nanoemulsion · Self-emulsification · Bioavailability · Bioactive

1 Introduction

Pharmaceutical dosage formulation is designed with an objective to deliver the effective concentration of drug to the site of action to obtain the desired therapeutic response. Oral drug delivery is generally preferred but is frequently dependent upon the bioavailability of the active form of the drug and the metabolism of the drug in the system. The oral delivery of hydrophobic candidate like drug of herbal origin presents a major challenge which is why almost 40–50% of drug compounds discovered are not successfully delivered through a specific dosage form as these new drug candidates display low solubility in water [1]. It ultimately leads to poor oral bioavailability, high inter-intra subject variability, and lack of dose proportionality [2]. Oral drug delivery has been a most fascinated route of drug administration for treatment of various ailments but it is impeded owing to undesirable characteristics of therapeutics such as high lipophilicity, poor aqueous solubility, frequent liver metabolism and high intra- inter subject variability, and lack of dose-dependent response. Dissolution process which is of great importance represents the solubility of drug and interns its absorption. It is low for lipophilic candidate like drug of herbal origin leading to its sub-therapeutic concentration in the in vivo system. Nowadays much attention has been focused on lipid-based drug delivery system, which involved the incorporation of therapeutic agent into lipidic system to improve the bioavailability of poorly soluble therapeutics [3]. Among the various lipid-based formulations, one of the most popular and commercially feasible formulation approach in resolving these problems is self-nanoemulsifying drug delivery system (SNEDDS). Their compatibility with a variety of molecular structures of bioactive phytochemicals, oils, surfactants, and co-solvents and ease of processing and manufacture make SNEDDS an attractive drug carrier for improving the efficacy of bioactive phytochemicals. Different marketed products of SNEDDS formulation of various therapeutic class are available and further validate its significance to solve out the biopharmaceutical challenges of poorly soluble bioactive phytochemicals/therapeutics after oral administration.

2 Self-Nanoemulsifying Drug Delivery System: an Overview

SNEDDS is defined as an isotropic mixture of natural or synthetic oils, solid or liquid surfactants, or alternatively, one or more hydrophilic solvents and co-solvents/surfactants which spontaneously emulsify when exposed to the fluids of the GIT to form oil-in-water nanoemulsions [3, 4]. It provides the advantage of increased drug loading capacity when compared with other lipid-based formulations such as emulsions, vesicular systems like liposomes, niosomes, transfersomes, or ethosomes, and solid lipid nanoparticles. The main rationale behind the design and development of SNEDDS is to improve the oral bioavailability of poorly bioavailable lipophilic therapeutics/bioactive phytochemicals. Such therapeutics have poor aqueous solubility and intermediate partition co-efficient values ($2 < \text{logp} < 4$) which hinder their absorption through biological membrane. Rapid release of the drug and increased drug solubilization in the gastro-intestinal lumen were suggested to be responsible for the improved drug bioavailability. The efficiency of emulsifiers in SNEDDS is commonly related to their ability to form fine droplet size of the emulsion on exposure to water, having polarity favoring faster rate of the drug release [5]. For therapeutic agent subject to dissolution-rate-limited absorption like drug of herbal origin, SNEDDS may improve the rate and extent of absorption, as well as the reproducibility of the blood level-time profile [6].One additional advantage in using SNEDDS with oral drugs that are P-gp substrates is their capacity to incorporate/or solubilize high levels of specific and nonspecific P-gp inhibitors [4, 6]. It is also reported that vegetable and animal oils used as formulation component in development of SNEDDS may have differential effects on particular hepatic CYP isoforms and may adjoin to the unevenness in metabolism when xenobiotics are delivered in lipid-based vehicle [5]. It has been used as a carrier for various lipophilic anticancer drugs of plant origin such as etoposides, paclitaxel, docetaxel, and camptothecin. SNEDDS can affect absorption of various therapeutics in many ways. It may act by enhancing drug solubilization in the intestinal surroundings by modifications to the composition and character of the colloidal environment. It may also improve the absorption by interacting with enterocyte based transport and metabolic processes, in that way potentially altering drug uptake, efflux, disposition, and the development of metabolites inside the enterocyte. In addition, it may improve by changing the pathway (portal vein versus intestinal lymphatic system) of drug transport to the systemic circulation i.e. which in turn can decrease first-pass drug metabolism as intestinal lymph moves straightforwardly to the systemic circulation without first passing through the liver. Some other factors, which modulate the biopharmaceutical properties of the therapeutics being a SNEDDS component should have good solvent capacity; chain length of fatty acid plays a significant role in efficient intestinal lymphatic transport to further improve the efficacy of therapeutics. In addition, formulation excipient selection (like lipids, surfactants) may also play critical role as it may affect efflux transporter function as well as metabolism within the enterocyte [4]. Therefore, certain SNEDDS composition may provide significant improvement in oral efficacy of poorly bioavailable bioactive phytochemicals besides solubilization benefits alone.

2.1 Component of SNEDDS

The major components present in SNEDDS formulation are oils, surfactant and co-surfactant or co-solvent. Oil phase of triglyceride vegetable oils has many advantages in lipid-based delivery systems. Oral absorption of lipophilic therapeutics through oily solution seems to be desirable because the digestion of triglycerides produces free fatty acids along with simpler glycerides which ultimately solubilized within bile salt micelles to form colloidal dispersion. Furthermore, it promotes the intestinal lymphatic transport of lipophilic therapeutics and favors increased absorption through GIT [7]. Modified or hydrolyzed vegetable oils have been widely used since these excipients form good emulsification system with a large number of surfactants approved for oral administration and exhibit better drug solubility properties [8]. Novel blends of oil like polyglycolyzed glycerides with polyethylene glycol have also been explored to solubilize the poorly soluble drug for greater bioavailability.

Surfactants are the surface active agents which reduce the intrafacial tension between two surfaces. Generally, SNEDDS include only non-ionic surfactants as they are less noxious than cationic and anionic surfactants. Non-ionic surfactants like tween, cremophor EL, and cremophore RH are relevant in oral formulations which are manifested by numerous successful marketed products [2]. Surfactants of natural origin are safer than synthetic surfactants [8]. Stable SNEDDS contain 30–60% surfactants. More than this range of concentration of surfactants causes GI irritation. Surfactant having intermediate HLB (8–12) value called water-insoluble surfactants may adsorb strongly at oil–water interface and form micelles. However, surfactants having HLB value $\geq$12 called water-soluble surfactants provide good self-emulsifying performance due to immediate formation of o/w droplets and/or rapid spreading of the formulation in the aqueous media. When a drug has insufficient solubility in either component of formulation to incorporate required therapeutic dose, co-solvent like polyethylene glycol, propylene glycol, and ethanol are used to increase the drug load in the formulation. To increase the drug load in formulation that drug should have very good solubility in co-solvents. However, co-solvent of volatile nature like ethanol has potential to evaporate into the shells of gelatin capsule leading to drug precipitation [8]. Sometimes lipid soluble antioxidant such as butylated hydroxytoluene (BHT), butylated hydroxyanisole (BHA), β-carotene, propyl gallate, or α-tocopherol, may be included in formulation to protect either unsaturated fatty acid chain used in formulation or drug from oxidation [9].

2.2 Evaluation of SNEDDS

The primary evaluation of SNEDDS is carried out by visual inspection. The rate of self-emulsification, droplet size distribution, and turbidity measurement give idea about the robustness of the formulation [7, 10]. The droplet size of the SNEDDS is a crucial factor in self-emulsification performance because it determines the rate and extent of drug release as well as absorption. Thermodynamic stability study was carried out to check physical stability of the SNEDDS under stress conditions. The formulations loaded with bioactive phytochemicals were subjected to stress tests, which comprise heating–cooling cycle (45 °C and 4 °C), freeze–thaw cycle (−21 °C and + 25 °C), and centrifugation tests (5000 rpm for 30 min) [11]. Physical stability was continuously monitored over the period of time. Various aspects like phase separation, creaming, cracking, drug precipitation, and turbidity were observed at room temperature. Dispersibility test of SNEDDS formulation involves the assessment of efficiency of self-emulsification of formulation using a dissolution apparatus. The in vitro performance of the formulations is visually assessed. Formulation with rapid self-emulsification (within 1 min) having clear and bluish appearance indicates formation of nanoemulsion on dispersion in GIT [11]. Turbidimetric analysis of SNEDDS is carried out to monitor the growth of emulsification. Self-emulsifying formulation is added to suitable medium under continuous stirring and rise in turbidity is determined using a turbidimeter. However, change in turbidity rate is quite impractical to measure due to the very short time required in complete emulsification of SNEDDS [12]. The droplet size distribution of the nanoemulsion formed after infinite dilution of SNEDDS to mimic the GIT dilution is determined by photon correlation spectroscopy using dynamic light scattering technique [12]. The technique is based on fluctuations in light scattering due to Brownian motion of the particles. Light scattering is monitored at 25 °C at 90° angle. In addition, content of therapeutics loaded in SNEDDS was determined by extracting the encapsulated therapeutics by dissolving in suitable solvent like methanol. Drug content in the solvent extract was analyzed by suitable analytical method against the standard solvent solution of drug.

2.3 Application of SNEDDS

SNEDDS is widely explored drug delivery approach for the oral administration of poorly soluble therapeutics that exhibit dissolution-rate-limited absorption. The droplets size in nano-dimension encapsulating poorly absorbable therapeutics provides large interfacial area from which therapeutics/bioactive phytochemicals can be quickly released and ultimately enhancing the efficacy of poorly water-soluble therapeutics after oral administration [4]. The increase in efficacy of therapeutics exploiting SNEDDS allowing the reduction in therapeutics dose and minimize drug-associated untoward effects. The oil phase of SNEDDS encapsulating the bioactive phytochemicals and protect the loaded therapeutics from hydrolysis and oxidation. SNEDDS is also known to exhibit pH independent absorption throughout intestinal mucosa thus nullifying the effect of pH on drug absorption through biological membranes [10]. In addition, reason for poor bioavailability of various therapeutics after oral administration is due to the substrate for intestinal cytochrome P450 and multidrug efflux pump. The excipients (surfactant, co-surfactant/co-solvent) used in formulation development of SNEDDS have property to modulate the P-gp and reported as efflux pump inhibitor [5]. The formulation design of SNEDDS involving excipients of P-gp inhibitor characteristics ultimately resulted in increase absorption of such therapeutic candidates.

3 SNEDDS for Improving Efficacy of Herbal Bioactive Compounds

SNEDDS are designed and evaluated for their proven efficacy to improve solubility or dissolution of poorly bioavailable drugs without compromising stability and cost of the formulation. This drug delivery system has been accepted industrially owing to lower process cost, high stability, reproducibility, and better patient compliance. Several self-emulsifying systems have been formulated and commercialized [2, 13]. This delivery system has been proven efficacious to improve the solubility or dissolution as well as efficacy of some poorly soluble herbal drugs. Various herbal bioactive compounds of different therapeutic class (such as hepatoprotective, anti-inflammatory, cardioactive, antimalarial, and anticancer agents, etc.) have been exploited to improve their efficacy through SNEDDS formulation design. Contemporary research in this area and their outcome details are discussed below.

3.1 Silymarin

Silymarin is a bioactive phytochemical of class flavonolignans that is extracted from the milk thistle *Silybum marianum* (L.) Gaertn. It has been widely reported for different therapeutic activities such as pharmacological properties like hepatoprotective, cardioprotective, anti-inflammatory, antioxidant, and anticancer effects. Although, its oral efficacy is limited due to the poor absorption, rapid metabolism, and ultimately poor oral bioavailability [14].Wu et al. demonstrated in his study that SNEDDS successfully improves the oral bioavailability of silymarin using tween 80, ethyl alcohol, and ethyl linoleate as formulation excipients of SNEDDS [15]. The mechanism responsible for improvement in oral bioavailability might be due to the enhanced lymphatic transport pathway through the SNEDDS formulation system.

3.2 Wurenchun

The ethanol extract of seeds and fruits of *Schisandra chinensis* which mainly contains schisandra lignans is commonly referred to as Wurenchun. It has been used in traditional system of Chinese medicine for their antihepatotoxic activity. Although, poor water-solubility of Schisandra lignans resulting in low bioavailability after oral administration [16]. Hancke et al. developed wurenchun containing SNEDDS to improve the solubility, dissolution rate, and oral bioavailability of wurenchun [17]. The formulation excipients used in the SNEDDS development were oleic acid, tween 20, and transcutol P which ultimately encapsulated the wurenchun in oil droplet of dimension approx. 240 nm upon dilution with distilled water. The SNEDDS formulation found to be significantly improved the oral bioavailability of Wurenchun compared to commercial capsule dosage forms.

3.3 Oleanolic Acid

Oleanolic acid is a pentacyclictriterpenoid phytochemical and possesses many significant therapeutic activities such as hepatoprotective, anti-inflammatory, anti-oxidant, and anticancer activities [18]. This bioactive phytochemical reported as poorly water-soluble which results in poor bioavailability after oral administration. To improve the oral delivery of oleanolic acid, SNEDDS formulation was developed by Xi et al. [19]. It was found that oleanolic acid exhibits 20 mg/g solubility in the composition of sefsol 218, cremophor EL, labrasol, and transcutol P. The oleanolic acid containing SNEDDS exhibited 2.4-fold increase in relative oral bioavailability compared to the commercial tablet.

3.4 Curcumin

Curcumin (CUR) is a polyphenol bioactive phytochemical obtained from the rhizome of the *Curcuma longa* [20]. This yellow colored polyphenolic compound has significant therapeutic activities against various disorders. It possesses antiangiogenesis, anticarcinogenic, anti-inflammatory and antimicrobial, and anti-rheumatic activities. In addition, curcumin also exhibited *neuroprotective*, cardioprotective, nephroprotective, and hypoglycemic activities. Although, poor water solubility of curcumin resulted in very low bioavailability after oral administration. The curcumin containing SNEDDS formulation was developed by Cui et al. [21]. It was found that curcumin exhibit 21 mg/g solubility in the composition of ethyl oleate, cremophor EL, and PEG 400. The dissolution study of curcumin containing SNEDDS observed for more than 95% of release of curcumin from SNEDDS within 20 min in pH 1.2 or pH 6.8 dissolution media. However, crude curcumin exhibited only <2% for drug release in 60 min. The curcumin containing SNEDDS exhibited significant improvement in oral absorption of curcumin compared to its suspension.

3.5 Ginkgolide

The ginkgolides are diterpenoids isolated from Ginkgo biloba L. and exhibit significant antagonist activity platelet-activating factor (PAF) [22]. The ginkgo extracts which contain ginkgolides as bioactive phytochemicals exhibited neuroprotective activity under different conditions like hypoxia/ischemia, seizure activity, and peripheral nerve damage. *Ginkgo biloba* extract was found to be improving the cognitive function of elderly people. The SNEDDS was developed by Tang et al. to improve absorption of sparingly soluble ginkgo biloba extract after oral administration [23]. The formulation excipients used in the SNEDDS development were ethyl oleate, 2-propanediol, and tween 80 along with cremophor EL which ultimately encapsulated the ginkgo biloba extract in oil droplet of dimension approx. 100 nm upon dilution with distilled water. The ginkgo biloba extract containing SNEDDS exhibited approx. 1.5-fold increase in relative oral bioavailability compared to the commercial tablet of ginkgo biloba extract. This investigation is proof of concept which validates efficient delivery of the sparingly soluble therapeutics (like *Ginkgo biloba* extract) exploiting SNEDDS formulation after oral administration.

3.6 Puerarin

Puerarin is an isoflavonoid obtained from the root of *Pueraria lobata.* It has been used to increase the blood circulation in cardiovascular and cerebrovascular disorders. Also, shown significant therapeutic activities in diabetes, osteonecrosis, Parkinson's disease, Alzheimer's disease, endometriosis, and cancer [24]. The puerarin containing SNEDDS was developed by Quan et al. to evaluate its absolute bioavailability exploiting oleic acid, propylene glycol, and tween 80 as formulation components [25]. The author has reported puerarin containing SNEDDS exhibited 24.8% bioavailability after oral administration in beagle dogs.

3.7 β-Artemether

β-Artemether is prodrug of dihydroartemisinin derived from *Artemisia annua.* This bioactive phytochemical exhibited significant antimalarial activity due to possessing an endoperoxide lactone ring [26]. The poor aqueous solubility of β-Artemether resulted in its low bioavailability after oral administration. The β-Artemether containing SNEDDS was developed by Mandawgade et al. using a novel, indigenous natural lipophile as an oily phase to improve the delivery of this antimalarial therapeutics [27]. The developed formulation system exhibited droplet size of <100 nm and is found to be a significant improvement in the antimalarial activity in mice model infected with lethal ANKA strain of Plasmodium berghei in comparison to the marketed formulation (Larither).

3.8 Camptothecin

Camptothecin is a pentacyclic alkaloid isolated from the stem wood of *Camptotheca acuminate* and its derivative approved for treatment of cancer [28]. It is a topoisomerase-I inhibitor but oral efficacy is low due to poor absorption and bioavailability. The SNEDDS formulation containing 9-nitrocamptothecin was formulated by Lu et al. with droplet size <50 nm [29]. The SNEDDS formulation containing 9-nitrocamptothecin exhibited significant antitumor activity after oral administration in nude mice bearing human ovarian cancer xenografts.

3.9 Oridonin

Oridonin is a diterpenoid obtained from *Rabdosia rubescens.* It exhibited significant therapeutic activity which includes antitumor, anti-inflammatory, antimicrobial, hepatic fibrosis prevention, and neurological effects [30]. The SNEDDS formulation containing oridonin was formulated by Zhang et al. with formulation composition of maisine 35–1, labrafac CC, cremophor EL, and transcutol P [31]. The in vitro release exhibited complete release of oridonin from SNEDDS in approx.12 h and showed 2.2-fold increase in relative bioavailability of oridonin compared to suspension after oral administration.

3.10 Thymoquinone

Thymoquinone is bioactive phytochemicals obtained from black seed oil (*Nigella sativa*). Literature reported promising therapeutic activities against various disorders. It exhibits antioxidant, anti-inflammatory, antibacterial, hepatoprotective, antimutagenic, and antitumor activities. It improves the anticancer activity of cytotoxic drugs and attenuates their side effects as well [32]. Its poorly water solubility is responsible factor for the low bioavailability after oral administration. The SNEDDS formulation containing thymoquinone was formulated by Kalam et al. with 3.87-fold increase in bioavailability of thymoquinone compared to suspension after oral administration [33]. The thymoquinone containing SNEDDS formulation exhibited significant hepatoprotective effect compared to suspension after in vivo oral administration.

3.11 Resveratrol

Resveratrol is a phytoalexin produced by several plants in response to injury or pathogen attack. It is a stilbenoid and found in the skin of grapes, blueberries, raspberries, mulberries, and peanuts. It has very high antioxidant potential and exhibits anti-inflammatory, anticarcinogenic, cardioprotective, vasorelaxant, phytoestrogenic, and neuroprotective activities [34]. The poor aqueous solubility, extensive pre-systemic metabolism, and occurrence of enterohepatic recirculation are major hurdle in its formulation development for oral application. The SNEDDS formulation containing resveratrol was formulated by Singh et al. with formulation composition lauroglycol FCC as lipid phase while labrasol and transcutol P used as surfactant and co-surfactant, respectively [35]. The developed formulation exhibited significant improvement in drug release profile of resveratrol compared to pure drug. In addition, developed SNEDDS formulation containing resveratrol have shown

4.31-fold increase in AUC compared to pure drug indicating remarkable enhancement in absorption of resveratrol after oral administration through SNEDDS.

3.12 Rutin

Rutin is a bioflavonoid obtained from *Ruta graveolens*. It consists of flavonol, quercetin, and the disaccharide rutinose as quercetin-3-O-rutinoside and sophorin. It is a citrus flavonoid widely distributed in wide variety of vegetables and fruits including apples, buckwheat, most citrus, figs, as well as both black and green tea. It has significant antioxidant properties and exhibited various therapeutic activities including cytoprotective, vasoprotective, neuroprotective, cardioprotective, and anticancer activities [36]. The SNEDDS formulation containing rutin was formulated by Singh et al. with formulation composition sefsol, vitamin E, solutol HS 15, and transcutol P [37]. The droplet size of the developed SNEDDS was found to be approx. 16 nm and demonstrated significant improvement in the percentage of drug release compared to suspension formulation of rutin. The in vivo performance of developed SNEDDS containing rutin exhibited 2.3-fold increase in relative oral bioavailability and helps in improving oxidative stress in different neurodegenerative disorders.

3.13 Lutein

Lutein is a naturally occurring carotenoids and synthesized by various green leafy vegetables like spinach, kale, and yellow carrots. It is a fat soluble antioxidant with significant anti-inflammatory properties and many beneficial activities, specifically on eye health. Also, improving cognitive function, minimizing cancer risk, and positive effect on cardiovascular health [38].

The SNEDDS formulation containing lutein was formulated by Yoo et al. and assesses its impact on bioavailability of warfarin [39]. The formulation composition of lutein SNEDDS was peceol, labrasol, transcutol-HP, and lutrol-E400 and complete dissolution is achieved after 15 min while commercial product remains not detectable. The SNEDDS formulation containing lutein is found to influence the bioavailability of warfarin by 10% increase in Cmax and AUC of the drug in Sprague Dawley rats.

3.14 Hesperetin

Hesperetin is a naturally occurring flavanone-glycoside found in citrus fruits such as lemons and sweet oranges. It possesses various therapeutic activities which include cholesterol-lowering effects, anti-inflammatory and anticancer activities, improved microcirculation, recovery of venous ulcers, inhibition of chronic venous insufficiency and hemorrhoids, and prevention of post-operative thromboembolism

[40]. The SNEDDS formulation containing hesperetin along with bicalutamide was formulated by Arya et al. to minimize drug induced toxicity of bicalutamide [41]. The droplet size of developed formulation system was found to be approx. 30 nm with nearly 90% of encapsulation efficiency for both the drugs. The developed formulation system significantly decreased the nephrotoxicity biochemical markers and exhibited significant increase in the AUC value of both drugs compared to its aqueous suspension counterpart in Sprague Dawley rats. Also, the author postulated that the developed SNEDDS formulation system absorbed through clathrin-mediated endocytosis and follow the lymphatic transport absorption pathway.

3.15 Berberine

Berberine is a benzylisoquinoline alkaloid found in the roots, rhizomes, stems, and bark of plant group called Berberis. It possesses various therapeutic activities which include reduction in blood sugar, improve heart health, and exhibited significant antimicrobial activity against several microbes [42]. The SNEDDS formulation containing berberine was formulated by Ke et al. with formulation composed of castor oil, tween 20, and glycerol [43]. The droplet size of developed formulation system was found to be approx. 60 nm. The optimized formulation demonstrated 90% berberine release in 20 min while commercial tablet and capsule dosage form of berberine required 2 h to release 90% of berberine. The results of developed SNEDDS of berberine promise and indicate a potential alternative approach for improving the oral absorption of poorly soluble bioactive phytochemicals.

3.16 Apigenin

Apigenin is a naturally occurring plant flavone which commonly occurs in fruits and vegetables. It is obtained from the wide range of plants including *Matricaria recutita*. It possesses remarkable therapeutic activities which include anti-inflammatory, antioxidant, and anticancer properties. Also, it exhibited antiviral and antibacterial properties and minimizes the blood pressure as well [44]. The SNEDDS formulation containing apigenin was formulated by Zhao et al. to enhance the solubility and dissolution rate [45]. The developed SNEDDS formulation is composed of cremophor EL, transcutol-HP, and capryol 90 with droplet size approx. 17 nm and spherical in appearance. The release of apigenin from developed SNEDDS was achieved about 95% in 10 min. Authors suggested that SNEDDS would be a potential carrier to enhance the oral absorption of poorly water-soluble bioactive.

3.17 Daidzein

Daidzein is an isoflavone found exclusively in soybeans and other legumes. It possesses the therapeutic activities to modulate the oxidative stress, angiogenesis, cell growth, and apoptosis. Also, observe for reduced incidence of breast and prostate cancers, cardiovascular disease, osteoporosis as well as cause reduction in cholesterol level in blood [46]. The SNEDDS formulation containing daidzein was formulated by Shen et al. with formulation composed of ethyl oleate, cremophor RH 40, and polyethylene glycol 400 [47]. The results obtained from developed formulation system indicate that dissolution rate of daidzein from SNEDDS was remarkably more than the conventional tablet. Furthermore, relative bioavailability of developed SNEDDS was improved about 2.5-fold compared to control. The obtained results suggest that SNEDDS technique provides a potential way to improve the oral efficacy of poorly soluble and bioavailable bioactive therapeutics/phytochemicals.

3.18 Withaferin A

Withaferin A is a bioactive phytochemical obtained from the leaves of *Withania somnifera*. It possesses various therapeutic activities which include anti-inflammatory, antistress, antioxidant, immunomodulatory, antiangiogenesis, and anticancer activities [48]. The SNEDDS formulation containing extract of *Withania somnifera* consist of Withaferin A was formulated by Gohel et al. with formulation composition of capmul MCM, tween 20, and ethanol [49]. The developed SNEDDS formulation found to be demonstrated higher drug release compared to the pure extract of *Withania somnifera*.

Outcome details of SNEEDS technique to improve the bioavailability/efficacy of different herbal bioactive compounds are summarized in Table 1.

4 Conclusion

SNEDDS are considered as a promising approach to improve the efficacy of poorly aqueous soluble drugs of herbal origin. However, the commercialization of this technique has been limited to a few marketed formulations, the reason perhaps could be the lack of in vitro tests depicting the in vivo performance of the formulations.

Conflict of Interest Nil.

Table 1 Improving efficacy of bioactive phytochemicals through SNEDDS

S. No.	Phytochemicals	Class	Therapeutic use	Formulation component	Observation	Ref.
1.	Silymarin	Flavonolignans	Hepatoprotective	Ethyl linoleate, tween 80, and ethyl alcohol	Approx. 49-fold increase in bioavailability of silymarin through SNEDDS	[15]
2.	Wurenchun	Schisandra lignans	Hepatoprotective	Oleic acid, tween 20, and transcutol P	Approx. 3-fold increase in bioavailability of wurenchun through SNEDDS	[17]
3.	Oleanolic acid	Pentacyclictriterpenoid	Hepatoprotective and antioxidant	Sefsol 218, cremophor EL, and labrasol	Approx.2.4-fold increase in bioavailability of oleanolic acid through SNEDDS	[19]
4.	Curcumin	Polyphenol	Anti-inflammatory	Ethyl oleate, cremophor EL, and PEG 400	Oral absorption improve 3.86 times higher through SNEDDS	[21]
5.	Ginkgolide	Diterpenoids	Cognitive enhancer	Ethyl oleate, tween 80, cremophor EL, and 1, 2-propanediol	Approx. 1.5-fold increase in bioavailability of ginkgolide through SNEDDS	[23]
6.	Puerarin	Isoflavonoid	Cardioactive	Oleic acid, tween 80, and propylene glycol	Improved absorption through SNEDDS and 24.8% bioavailability after oral administration	[25]
7.	B-Artemether	Sesquiterpene lactones	Antimalarial	Capryol 90, plurol oleique CC 97, tween 80, and gelucire 44/14	Significant improvement in antimalarial activity through SNEDDS	[27]
8.	Camptothecin	Pentacyclic alkaloid	Anticancer	Ethyl oleate, tween 80, PEG 400, and ethanol	Approx. 2.2-fold increase in bioavailability of camptothecin through SNEDDS	[29]
9.	Oridonin	Diterpenoid	Anticancer	Maisine 35–1, labrafac CC, cremophor EL, and transcutol P	Approx. 2.2-fold increase in bioavailability of oridonin through SNEDDS	[31]
10.	Thymoquinone	*Terpene*	Hepatoprotective	Capryol-90, tween-20, and isopropyl alcohol	Approx. 3.87-fold increase in bioavailability of oridonin through SNEDDS	[33]

(continued)

Table 1 (continued)

S. No.	Phytochemicals	Class	Therapeutic use	Formulation component	Observation	Ref.
11.	Resveratrol	Stilbenoid	Antioxidant and anticancer	Lauroglycol FCC, labrasol, and transcutol	Approx. 4.31-fold increase in bioavailability of oridonin through SNEDDS	[35]
12.	Rutin	Bioflavonoid	Antioxidant	Sefsol, vitamin E, solutol HS 15, and transcutol P	Approx. 2.3- fold increase in bioavailability of oridonin through SNEDDS	[37]
13.	Lutein	Carotenoids	Antioxidant	Peceol, labrasol, transcutol-HP, and lutrol-E400	Influence the bioavailability of warfarin by 10% increase in Cmax and AUC	[39]
14.	Hesperetin	Flavanone-glycoside	Hypolipidemic	Castor oil, cremophor EL, and triton X-10	Significantly decreased the nephrotoxicity biochemical markers and exhibited significant increase in the AUC	[41]
15.	Berberine	Benzylisoquinoline alkaloids	Antimicrobial	Castor oil, tween 20, and glycerol	Significant increase in release of berberine through SNEDDS	[43]
16.	Apigenin	Flavone	Anti-inflammatory	Cremophor EL, transcutol-HP, and capryol 90	Significant increase in release of apigenin through SNEDDS	[45]
17.	Daidzein	Isoflavones	Anticancer	Ethyl oleate, cremophor RH 40, and polyethylene glycol 400	Approx. 2.5-fold increase in bioavailability of daidzein through SNEDDS	[47]
18.	Withaferin A	Steroidal lactone	Immunomodulator	Capmul MCM, tween 20, and ethanol	Demonstrated higher drug release through SNEDDS	[49]

References

1. Taha E, Ghorab D, Zaghloul A (2007) Bioavailability assessment of vitamin A self-nanoemulsified drug delivery systems in rats: a comparative study. Med Princ Pract 16:355–359
2. Tang B, Cheng G, Gu JC, Xu CH (2008) Development of solid self-emulsifying drug delivery systems: preparation technique and dosage form. Drug Dis Today 13:606–612
3. Shah NH, Carvajal MT, Patel CI, Infeld MH, Malick AW (1994) Selfemulsifying drug delivery systems (SEDDS) with polyglycolized glycerides for improving in vitro dissolution and oral absorption of lipophilic drugs. Int J Pharm. 106:15–23
4. Pouton CW (2006) Formulation of poorly water-soluble drugs for oral administration: physicochemical and physiological issues and the lipid formulation classification system. Eur J Pharm Sci 29:278–287
5. Charman SA, Charman WN, Rogge MC, Wilson TD, Dutko FJ, Pouton CW (1992) Self-emulsifying drug delivery systems: formulation and biopharmaceutic evaluation of an investigational lipophilic compound. Pharm Res 9:87–93
6. Pouton CW (1985) Self-emulsifying drug delivery systems: assessment of the efficiency of emulsification. Int J Pharm 27:335–348
7. Gershanik T, Benita S (2000) Self-dispersing lipid formulations for improving oral absorption of lipophilic drugs. Eur J Pharm Sci Biopharm 50:179–188
8. Constantinides PP (1995) Lipid microemulsion for improving drugs dissolution and oral absorption: physical and biopharmaceutical aspects. Pharm Res 12:1561–1572
9. Gibson L (2007) Lipid-based excipients for oral drug delivery. In: Hauss DJ (ed) Oral lipid-based formulations: enhancing the bioavailability of poorly water soluble drugs. Informa Healthcare Inc, New York, pp 33–61
10. Ahmad J, Kohli K, Mir SR, Amin S (2012) Self-emulsifying nano carriers for improved oral bioavailability of lipophilic drugs. Rev Adv Sci Eng 1(2):134–147
11. Shafiq S, Faiyaz S, Sushma T, Farhan JA, Khar RK, Mushir A (2007) Development and bioavailability assessment of ramipril nanoemulsion formulation. Eur J Pharm Biopharm 66:227–243
12. Patil P, Vandana P, Paradkar P (2007) Formulation of self-emulsifying drug delivery system for oral delivery of simvastatin: in vitro and in vivo evaluation. Actapharma 57:111–122
13. Porter CJH, Pouton CW, Liune JF, Charman WN (2008) Enhancing intestinal drug solubilisation using lipid based delivery system. Adv Drug Deli Rev 60:673–691
14. Javed S, Kohli K, Ali M (2011 Sep 1) Reassessing bioavailability of silymarin. Altern Med Rev 16(3):239
15. Wu W, Yang W, Que L (2006) Enhanced bioavailability of silymarin by self-microemulsifying drug delivery system. Eur J Pharm Biopharm. 63:288–294
16. Hancke JL, Burgos RA, Ahumada F (1999) Schisandra chinensis (Turcz) Baill. Fitoterapia 70:451–471
17. Shao B, Tang J, Ji H, Liu H, Liu Y, Zhu D, Wu L (2010) Enhanced oral bioavailability of Wurenchun (Fructus Schisandrae Chinensis extracts) by self-emulsifying drug delivery systems. Drug Dev Ind Pharm 36(11):1356–1363
18. Pollier J, Goossens A (2012) Oleanolicacid. Phytochemistry 77:10–15
19. Xi J, Chang Q, Chan CK, Meng ZY, Wang GN, Sun JB, Wang YT, Tong HHY, Ying Z (2009) Formulation development and bioavailability evaluation of a self-nanoemulsified drug delivery System of oleanolic acid. AAPS PharmSciTech 10(1):172–182
20. Ahmad MZ, Akhter S, Mohsin N, Abdel-Wahab BA, Ahmad J, Warsi MH, Rahman M, Mallick N (2014) Ahmad FJ1. Transformation of curcumin from food additive to multifunctional medicine: nanotechnology bridging the gap. Curr Drug Discov Technol 11(3):197–213
21. Cui J, Yu B, Zhao Y, Zhu W, Li H, Lou H, Zhai G (2009) Enhancement of oral absorption of curcumin by self-microemulsifying drug delivery systems. Int J Pharm 371:148–155
22. Omar SH (2013) Ginkgolides and neuroprotective effects. In: Ramawat K, Mérillon JM (eds) Natural products. Springer, Berlin, pp 3697–3741

23. Tang J, Sun J, Cui F, Zhang T, Liu X, He Z (2008) Self-emulsifying drug delivery systems for improving oral absorption of Ginkgo Biloba extracts. Drug Deliv 15(8):477–484
24. Zhou YX, Zhang H, Peng C (2014) Puerarin: a review of pharmacological effects. Phytother Res. 28(7):961–975
25. Quan D, Xu Gand WX (2007) Studies on preparation and absolute bioavailability of a self-emulsifying system containing Puerarin. Chem Pharm Bull 55(5):800–803
26. Wang J, Xu C, Wong YK, Li Y, Liao F, Jiang T, Tu Y (2019) Artemisinin, the magic drug discovered from traditional Chinese medicine. Engineering 5(1):32–39
27. Mandawgade SD, Sharma S, Pathak S, Patravale VB (2008) Development of SMEDDS using natural lipophile: application to β-Artemether delivery. Int J Pharm 362:179–183
28. Li F, Jiang T, Li Q, Ling X (2017) Camptothecin (CPT) and its derivatives are known to target topoisomerase I (Top1) as their mechanism of action: did we miss something in CPT analogue molecular targets for treating human disease such as cancer? Am J Cancer Res. 7(12):2350
29. Lu J, Wang J, Zhao S, Liu X, Zhao H, Zhang X, Zhou S, Zhang Q (2008) Self-microemulsifying drug delivery system (SMEDDS) improves anticancer effect of oral 9-nitrocamptothecin on human cancer xenografts in nude mice. Eur J Pharm Biopharm 69:899–907
30. Xu J, Wold EA, Ding Y, Shen Q, Zhou J (2018) Therapeutic potential of oridonin and its analogs: from anticancer and antiinflammation to neuroprotection. Molecules 23(2):474
31. Zhang P, Liu Y, Feng N, Xu J (2008) Preparation and evaluation of self-microemulsifying drug delivery system of oridonin. Int J Pharm 355:269–276
32. Barkat MA, Harshita AJ, Khan MA, Beg S, Ahmad FJ (2018) Insights into the targeting potential of thymoquinone for therapeutic intervention against triple-negative breast cancer. Curr Drug Targets 19(1):70–80
33. Kalam MA, Raish M, Ahmed A, Alkharfy KM, Mohsin K, Alshamsan A, Al-Jenoobi FI, Al-Mohizea AM, Shakeel F (2017) Oral bioavailability enhancement and hepatoprotective effects of thymoquinone by self-nanoemulsifying drug delivery system. Mater SciEng C Mater Biol Appl. 76:319–329
34. Salehi B, Mishra AP, Nigam M, Sener B, Kilic M, Sharifi-Rad M, Fokou PV, Martins N, Sharifi-Rad J (2018) Resveratrol: a double-edged sword in health benefits. Biomedicine 6(3):91
35. Singh G, Pai RS (2015) Trans-resveratrol self-nano-emulsifying drug delivery system (SNEDDS) with enhanced bioavailability potential: optimization, pharmacokinetics and in situ single pass intestinal perfusion (SPIP) studies. Drug Deliv 22(4):522–530
36. Sharma S, Narang JK, Ali J, Baboota S (2016) Synergistic antioxidant action of vitamin E and rutin SNEDDS in ameliorating oxidative stress in a Parkinson's disease model. Nanotechnology 27(37):375101
37. Buscemi S, Corleo D, Di Pace F, Petroni ML, Satriano A, Marchesini G (2018) The effect of lutein on eye and extra-eye health. Nutrients 10(9):1321
38. Yoo J, Baskaran R, Yoo BK (2013) Self-nanoemulsifying drug delivery system of lutein: physicochemical properties and effect on bioavailability of warfarin. Biomol Ther 21(2):173
39. Sangpheak W, Kicuntod J, Schuster R, Rungrotmongkol T, Wolschann P, Kungwan N, Viernstein H, Mueller M, Pongsawasdi P (2015) Physical properties and biological activities of hesperetin and naringenin in complex with methylated β-cyclodextrin.Beilstein. J Org Chem 11:2763–2773
40. Arya A, Ahmad H, Tulsankar S, Agrawal S, Mittapelly N, Boda R, Bhatta RS, Mitra K, Dwivedi AK (2017) Bioflavonoid hesperetin overcome bicalutamide induced toxicity by co-delivery in novel SNEDDS formulations: optimization, in vivo evaluation and uptake mechanism. Mater Sci Eng C 71:954–964
41. Birdsall TC, Kelly GS (1997) Berberine: therapeutic potential of an alkaloid found in several medicinal plants. Alt Med Rev 2(2):94–103
42. Ke Z, Zhu Z, Xu Z, Fang C, Hu S (2015) Formulation design and in vitro evaluation of berberine-loaded self-nanoemulsifying drug delivery system. Trop J Pharm Res 14(5):747–752
43. Shukla S, Gupta S (2010) Apigenin: a promising molecule for cancer prevention. Pharm Res 27 (6):962–978

44. Zhao L, Zhang L, Meng L, Wang J, Zhai G (2013) Design and evaluation of a self-microemulsifying drug delivery system for apigenin. Drug Dev Ind Pharm 39(5):662–669
45. Lamartiniere CA, Wang J, Smith-Johnson M, Eltoum IE (2002) Daidzein: bioavailability, potential for reproductive toxicity, and breast cancer chemoprevention in female rats. Toxicol Sci 65(2):228–238
46. Shen Q, Li X, Yuan D, Jia W (2010) Enhanced oral bioavailability of daidzein by self-microemulsifying drug delivery system. Chem Pharm Bull 58(5):639–643
47. Dutta R, Khalil R, Green R, Mohapatra SS, Mohapatra S (2019) Withania somnifera (Ashwagandha) and withaferin A: potential in integrative oncology. Int J Mol Sci 20(21):5310
48. Gohel M, Kureshi R, Hingorani L, Patel A, Thakkar V, Baldaniya L, Gandhi T (2015) Formulation and evaluation of Self-nano emulsifying Drug Delivery System For solubility and bioavailability Enhancement of poorly water soluble Withaferin A. Indian Drugs 52 (10):49–54
49. Ganeshpurkar A, Saluja AK (2017) The pharmacological potential of rutin. Saudi Pharm J. 25 (2):149–164

Potential of Nano-Structured Drug Delivery System for Phytomedicine Delivery

Vineet Kumar Rai, Ghanshyam Das Gupta, Faheem Hyder Pottoo, and Md. Abul Barkat

Abstract

Based on the problems associated with the delivery of herbal medicines, various investigations have been carried out. The field of formulation and development appeared progressive when we see the reports published in the previous few decades. Recent advancements in this field are based on the necessities of improved solubility, release, biological absorption, targetability, safety, dosing schedule, stability, efficacy, and patient compliance along with the decreased dose and dose-dependent side effects. Formulation scientists are now interested in the exploitation of new drug delivery system (NDDS) to deliver phytomedicine at the required rate for the entire treatment period and to channelize the phytomolecules at the desired site of action. These systems are meant to meet the unmet medical conditions that cannot be met using conventional dosage forms. Nanocarriers such as nanoparticles, nanocapsules, lipid carriers, solid lipid nanoparticles, phytosomes, nanoemulsion, and many more have shown an array of advantages while delivering phytomolecules to the site of action. Due to the proven advantages over conventional dosage forms, nano-phytomedicines have immense potential in the improvement of deliverability of the isolated and established herbal molecules. This chapter, however, includes various aspects of justification for the immense potential of the nano-structured drug delivery system for phytomedicine delivery and future prospects. In addition, the area of nanostructured carrier systems in the potential delivery of phytomedicine, and the

V. K. Rai (✉) · G. D. Gupta
Department of Pharmaceutics, ISF College of Pharmacy, Moga, Punjab, India

F. H. Pottoo
Department of Pharmacology, College of Clinical Pharmacy, Imam Abdulrahman Bin Faisal University, Dammam, Saudi Arabia

M. A. Barkat
Department of Pharmaceutics, College of Pharmacy, University of Hafr Al-Batin, Hafr Al-Batin, Saudi Arabia

S. Beg et al. (eds.), *Nanophytomedicine*,
https://doi.org/10.1007/978-981-15-4909-0_6

nano-phytomedicines which have been prepared for improved solubility, pharmacokinetic profile (ADME), bioavailability, targetability, efficacy, and safety are also highlighted in this chapter.

Keywords

New drug delivery system (NDDS) · Phytomedicine · Nanoparticles · Nanocapsules · Lipid carriers · Solid lipid nanoparticles · Phytosomes · Nanoemulsion

1 Introduction

The area of nano-structured drug delivery system has now well recognized. Literature proves that these carriers have immense potential to transform the current scenario of phytoactives based drug delivery approaches [1]. In lieu of that and array of investigations have been carried out reporting the modulation of physicochemical and biopharmaceutical characteristics of the drugs. Figure 1 highlights the vast potential of the nanostructured systems developed for phytomedicine delivery. Recent advancements in the field of phytomedicine delivery with the help of nanostructured carrier system are based on the necessities of improving the aqueous solubility of the poorly soluble phytoactives such as curcumin, andrographolide [2], etc., improving the release of the herbal molecules following oral route of administration, improving biological absorption through GI tract, improving targetability by avoiding non-specific drug availability to the unwanted site, improving safety [3], simplifying dosing schedule of the frequently prescribed drugs, improving stability of the phytoactives during extraction process, formulation development, and storage, improving therapeutic efficacy and hence patient compliance along with the decreased dose and dose-dependent side effects [4, 5]. As these systems are developed to meet the unmet medical conditions that could not be met using conventional dosage forms, nanocarriers such as nanoparticles, nanocapsules, lipid carriers, solid lipid nanoparticles [6], phytosomes, nanoemulsion [5], and many more have shown an array of advantages while delivering phytomolecules to the site of action. Due to the proven advantages over conventional dosage forms, nano-phytomedicines have immense potential in the improvement of deliverability of the isolated and established herbal molecules just by improving the solubility, pharmacokinetic profile (ADME), bioavailability, targetability, efficacy, and safety.

2 Nano-Structured Drug Delivery System for Improved Aqueous Solubility of the Phytomolecules

Solubility of the phytomolecules is one of the critical problems that limits their wide application. Most of the phytomolecules are the results of bioactivity guided fractionation and isolation process. Oral or parenteral use of the extracts or molecules other than aqueous extract are limited. Whenever direct availability of these molecules in

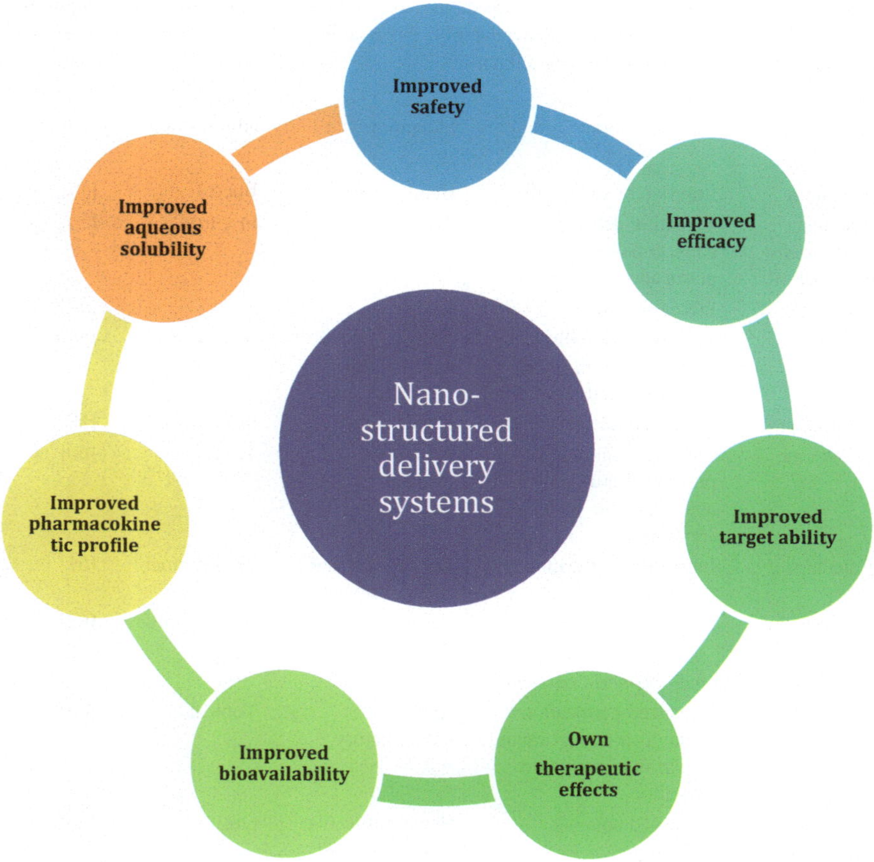

Fig. 1 Advantages of nano-structured system developed for phytomedicine delivery

the systemic circulation is required to maintain blood therapeutic level, additional measures have been taken to enhance the solubility inside the biological systems. The most usual practices while improving the solubility of the molecules are reduction in particle size, making solid dispersion and inclusion complexation, addition of co-solvent, addition of surfactants/solubilizers, etc [7, 8]. Nanopharmaceuticals are also reported to enhance the solubility of phytomolecules. Nanocarriers are novel means of solubility enhancement due to their ability in the formation of stable particles and droplets where drugs can easily be encapsulated or emulsified for the modulation of their physiochemical properties [9, 10]. These have vast potential in the enhancement of solubility of phytomolecules, for example, curcumin nanoparticles have increased its solubility by several folds [11]. Drugs of BCS class 2 are transformed into more soluble nano-phytomedicines so that maximum absorption through oral route can be ensured. As solubility is the rate-limiting step in such types of drugs, various measures have been taken just to increase their solubility so as to ensure a high level of drug into systemic circulation (Table 1). Modifying surface

Table 1 List of lipid based nano-phytomedicines prepared for the improvement of solubility, pharmacokinetic profile (ADME), bioavailability, target ability, efficacy, and safety along with nano-phytomedicines with therapeutic effects

Phyto-constituent	Value addition	Indication	Route of administration	Reference
Liposomes				
Silymarin	Improve bioavailability	Hepatoprotective	Buccal	[55]
Artemisia arborescens essential oil	Targeting delivery to cells, enhanced permeation	Antiviral	In vitro	[56]
Garlicin	Increased efficacy	–	–	[57]
Usnea acid	Increased solubility, Improved localization and prolonged release	Anti-mycobacterial	In vitro	[58]
Wogonin	Sustained release effect	Anticancer	In vivo	[59]
Colchicine	Enhance skin deposition, prolonged drug release and improved site specificity	Antigout	Topical	[60]
Catechins	Increased permeation through skin	Antioxidant and chemo-preventive	Transdermal	[61]
Breviscapin	Sustained drug delivery	Cardiovascular diseases	Intramuscular	[62]
Nano-structured lipid/polymeric carriers				
Triptolide	Enhanced permeation through stratum corneum by virtue of increased hydration	Anti-inflammatory	Topical	[63]
Flavonoids and lignans	Improved aqueous solubility	Hepatoprotective and antioxidant effects	Oral	[64]
Triptolide	Improved safety	Anti-inflammatory	Oral	[65]
Artemisinin	Sustained drug release	Anticancer	In vitro	[66]
Breviscapine	Prolong the half-life and decreased RES uptake	Cardiovascular and cerebrovascular	Intravenous	[67]
Camptothecin	Prolonged blood circulation and high deposition in tumors	Anticancer	In vitro	[68]
Phytosomes				
Silybin Flavonoids	Increased oral absorption of silybin	Hepatoprotective, antioxidant	Oral	[69]
Naringenin	Prolonged duration of action	Antioxidant activity	Oral	[70]

with polymers or lipids, making nanosized particles, increasing surface area are some of the measures taken for solubility enhancement most often [11, 12]. One of the investigations carried out by Piazzini et al. [13] reveals the utility of nanoemulsion in improving solubility of Vitex agnus-castus (VAC) extract along with improved permeability through a biological membrane. This approach was taken to improve the oral delivery of VAC extract. Triacetin, labrasol, cremophor EL, and water as oil, surfactant, co-surfactant, and aqueous phase, respectively, resulted in the reduction of the size of the drug-encapsulated droplets, i.e. up to 11.82 ± 0.125 nm and hence tenfold improvement in the VAC solubility. This technique led to the complete solubilization of the VAC extract in water. As nanoemulsion showed good potential in solubilizing VAC extract, the improved oral bioavailability of this extract was also expected to be improved due to improvement in both solubility and as well as permeability [13]. Nanoparticle systems are known to modify the hydrophobic surface of the phytoconstituents. As far as solubility improvement is concerned, nanoparticles approach has widely been investigated to improve the aqueous solubility of hydrophobic drugs. Nanoparticles are meant to increase the distribution of the drug due to which the solubility of the drug enhanced considerably. Combination of particle size reduction and surface modification worked wonders in many cases [14, 15]. In one of the examples, nanoparticles have been used to encapsulate two well-known polyphenols, i.e. morin and epicatechin in albumin nanostructures. Morin and epicatechin incorporated human serum albumin nanoparticles with nanosize particles ($\sim$$170 \pm 6$ nm to $\sim$$200 \pm 12$ nm) modified the surface in a manner in which drug distribution in the biological systems was improved considerably [12]. Solubility is not the sole parameter. It is always associated with drug dissolution, distribution, local availability, and bioavailability. Therefore, it is not investigated solely, however, in combination with other parameters like permeation, absorption, and bioavailability.

3 Nano-structured Drug Delivery System for Improved Pharmacokinetic Profile of Phytomolecules

Pharmacokinetics of an active moiety describes how the body affects its biopharmaceutical properties following absorption, distribution, metabolism, and excretion from the body after administration. When the drug enters into a biological environment of GIT, it follows the number of phases described by the acronym LADME. Liberation is another term, which is used today. It means the liberation of the active substance from the delivery system. It is also equally important as absorption. Most of the pharmacologist sites liberation and absorption of the drug as one phase. They consider the drug is often administered in an active form and following no release phase out there before absorption. However, phenomenon absorption is completely meaningless unless the liberation/release of the drug from novel carrier systems is justified. As far as nano-phytomedicine is concerned, liberation step should be justified firstly before making any pharmacokinetic profile of the dosage forms. One more acronym is used, i.e. ADME-Tox or ADMET that reflect the drug's toxicological aspect. Most of nano-phytomedicine meant to be sustained or controlled release of the drug. Controlling drug release prolongs drug staying inside the body and hence greater tissue distribution and redistribution are observed in case of cytotoxic drugs. That may lead to the death of normal cells of the body. Therefore, it is another important aspect that is looked into.

Availability of the active drug in solubilized form, thickness of the mucus membrane, log P, and many more factors affect the absorption of the drug through the biological membrane. Distribution of drug in the systemic circulation and organs depends on plasma protein binding of the drug, biological half-life, stability of the compound at specific pH or temperature, etc. Metabolic enzymes like cytochrome P450, glucuronosyltransferase enzymes may facilitate or hamper drug metabolism. The rate of excretion depends on the availability of metabolized drug and volume of distribution in the systemic circulation. Pharmacokinetic properties get affected following different routes of administration and the amount of drug administered. Nano-phytomedicines are prepared to alter the pharmacokinetic pattern of the actives of choice. Encapsulation directly involves in surface modification of the drug and hence modifies the rate of absorption of the less absorbed actives. One such investigation, i.e. carried out by Piazzini et al. reveals the utility of nanoemulsion in improving solubility because of which the permeability of VAC extract also improved when assayed using parallel artificial membrane permeation assay (PAMPA). Increased permeation of VAC when transformed in the form of nanoemulsion could improve oral bioavailability [13]. Absorption is directly associated with the permeation of actives through the biological membrane. Permeation is sometimes modulated to control the rate of absorption through skin also. This makes the formulation able to cross the skin following the transdermal route [16–18]. In a report, Lin et al. highlight the utility of ultra-deformable lipid (UDL) vesicles in expanding the application of Chinese herbal medicine, i.e. Imperatorin (IMP) from transdermal route also. Making cationic UDLs of high entrapment efficiency (60.32% $\pm$ 2.82%) and smaller particle size (>100 nm) provided 3.45-folds higher penetration of IMP through the skin as compared to that of its suspension counterpart. Surface modification of IMP by cationic UDLs worked wonders in this case by facilitating higher availability in the different layer skin due to increased transdermal flux, prolonged release of IMP [19]. Making nanoemulsion improved intestinal absorption of phytoestrogenic diarylheptanoids. The main source of phytoestrogenic diarylheptanoids is *C. comosa* that have good applicability in gynecological disorders. Limitation associated with the absorption of *C. comosa* was highlighted and encountered by developing its nanoemulsion to support its widespread application. Hexane extract of *C. comosa,* i.e. majorly composed of (4E,6E)-1,7-diphenylhepta-4,6-dien-3-ol (DA1) and (6E)-1,7-diphenylhept-6-en-3-ol (DA2) was estimated after having emulsified in terms of intestinal absorptions. Nanosized droplets of the developed nanoemulsion along with uniform size distribution and negative surface charge were observed some of the physicochemically altered parameters that resulted enhanced drug absorption. These nanodroplets showed improved intestinal absorption rates, i.e. five- to tenfolds faster than that of pure oil [20].

Distribution is other pharmacokinetic parameters that should be ensured for the maintenance of therapeutic blood concentration. For better distribution, measures are taken to reduce plasma protein binding of the drug, extending biological half-life and improving stability of the compound at specific pH or temperature of biological fluids. As per the preclinical report of Chang et al., oral bioavailability and tissue distribution of silymarin both have been improved considerably after encapsulating it in liposome and ethosome. The pharmacokinetic profile of this drug was observed to be changed after the surface modification by liposome and ethosome. Ethanolic and lipidic vesicles have vitally been investigated for the same purpose previously. This has happened due to the increased distribution of the silymarin in systemic circulation due to the increased solubility in aqueous environment. Less aqueous solubility was the major reason behind silymarin's limited oral bioavailability that has been ensured after employing vesicular approaches as these approaches ultimately improved its therapeutic effectiveness [21]. One more investigation of this type has previously been carried out to improve the pharmacokinetics and organ distribution of icariin by making its PG-liposomes for i.p. injection in small animals. As per the preclinical investigations, spherical vesicles of ICA-PG-liposomes with average particle diameter of 182.4 nm, 92.6% entrapment efficiency displayed improved icariin absorption from the GI tract, delayed average duration of retention (MRT(0-t)), enhanced area under the curve (AUC(0-t)), and Cmax, i.e. maximum plasma concentration as compared to that of ICA-PG-solution. ICAPG-liposomes improved icariin distribution into spleen, liver, lung, kidney, heart, and brain as compared to that of its solution counterpart. Increased AUC(0-t) in most of the tissues (except lung of mice), decreased clearance, prolonged MRT(0-t) and biological half-life in all of the tissues (except brain) were the result of increased volume of distribution of icariin in tissues [22]. In the latest investigation carried out for baicalin, Li et al. again proved the potential of liposomal drug delivery systems in improving the pharmacokinetics profile and bio-distribution of this drug. Baicalin (BA) which is supposed to be the key component of the dry root extract of *Scutellaria baicalensis*, has been used conventionally in the therapy of cerebral ischemia in Chinese clinics, and is neuroprotective in nature. Owing to its poor lipid solubility, its oral bioavailability is very low. In order to widen its clinical application, it was developed in the form of a novel liposomal carrier (BA-LP). It enhanced drug lipophilicity and hence the drug level in the brain owing to the enhanced lipophilicity. BA-LP had a particle size range of 160–190 nm, zeta potential of +5.7 mV, and 42% entrapment efficiency. Due to the controlled release properties, the in vitro release kinetic model of the drug found best fit to BA-LP with the biphasic dynamic model equation. Higher Cmax and AUC0–t, i.e. up to 1.5 to 2-folds were observed as compared to that of normal rats. That indicates higher distribution of BA in the brain as compared to the normal rats. Altered tissue distribution after making liposome led to the increased level of BA in heart, liver, spleen, lungs, and brain and decreased concentrations of BA in kidneys. Liposomal drug delivery systems found a greater place in delivering BA and hence BA-LP was considered as a potential carrier in improving therapeutic efficacy of BA like molecules [23]. Lipid based nanocarriers described potentially better in improving tissue distribution and pharmacokinetic profile of the drugs as compared to that of other nanocarriers developed for phytomolecules. It was proved by another example where docetaxel pharmacokinetics and tissue distribution improved when loaded in liposomes. It was mediated via novel galactosylated cholesterol derivatives synthesized with the help of lipase-catalyzed esterification in the organic part. Luo et al. synthesized galactosylated cholesterol derivative, cholesterol-diethenyl decanedioate-lactitol (CHS-DD-LA) via lipase-catalyzed esterification which improved pharmacokinetics and bio-distribution of docetaxel (DOC) liposomes after getting actively gathered in the liver as compared to the traditional DOC liposomes (DOC-L) and marketed dosage forms of DOC injection (DOC-i). Intravenous administration (2.5 mg/kg BW) in rats, decreased AUC, MRT and increased CL (p50.05) in the G-DOC-L group compared with DOC-L. The organ distribution of G-DOC-L was found thoroughly different from DOC-L. G-DOC-L showed outstanding potential in site-specific delivery to the liver that helped improve the pharmacological efficacy in hepatic malignancies [24]. As far as improvement in absorption and pharmacokinetic profile are concerned, nanostructured liposomal drug delivery systems had shown vast potential. Their ability in facilitating drug partitioning from the aqueous phase to lipid phase, in encapsulating both hydrophilic and lipophilic drug, in surface modification and their compatibility with the biological environment have been investigated well using plant based molecules. Importance of distribution in drug delivery is equally important as it ensures the higher bioavailability of the drugs. Tables 1 and 2 highlight some of the nanocarriers prepared for improved pharmacokinetic parameters of the drugs.

4 Nano-structured Drug Delivery System for Improved Bioavailability

Reports are available on various efforts made by the experts in enhancing stability as well as solubility of the herbal drugs by making their nanocarrier systems. However, bioavailability depends on other factors also, i.e. drug solubilization at blood pH, absorption of the drug, distribution of the solubilized drug into systemic circulation, first-pass metabolism, GI stability, and molecular weight of the drugs. Some of the approaches of bioavailability improvement in phytomedicine are followed by increasing absorption of the drug. In an investigation, silymarin solubility has been increased by making nanoformulations, which ultimately improved drug's bioavailability on account of which its therapeutic efficacy as a hepatoprotectant, antioxidant, antiviral, and anticancer agent have been improved. With the advancement of the technologies for the solubility enhancement, we have now a wide range of nanocarriers to entrap poorly water-soluble active plant ingredients. Most advanced techniques out of all are preparing nanovesicle and nanocarriers. Silymarin in aqueous nanovesicles, for instance, could be an efficient way of facilitating passive diffusion [25]. Moreover, SNEDDSs have been developed to increase the bioavailability after improving the solubility of the cyclovirobuxine D. The final preparation of SNEDDS was prepared by taking cyclovirobuxine D, oleic acid, Solutol SH15,

Table 2 List of nanoemulsions, vesicular systems, and polymeric particles developed for the improved solubility, pharmacokinetic profile (ADME), bioavailability, target ability, efficacy, and safety along with the therapeutic effects of phytomedicines

Nanoemulsions				
Zedoary turmeric oil	Improved aqueous dispersibility, stability and oral bioavailability	Hepatoprotection anticancer and antibacterial	Oral	[71]
Triptolide	Enhanced penetration of drugs through the stratum corneum by increased hydration	Anti-inflammatory	Topical	[63]
Salicylic acid	Controlled release and improved dermal retention	Anti-inflammatory activity	Topical	[4]
Mentha essential oil	Control the release, avoid the contact with the skin and reduce the evaporation	Anti-fungal activity against vaginal candidiasis	Topical	[5]
Andrographolide	Improved aqueous solubility, gastric stability and oral bioavailability	Hepatoprotective activity	Oral	[2]
Vesicular systems				
Colchicine	Increase skin penetration	Antigout	In vitro	[72]
Matrine	Improved the percutaneous permeation	Anti-inflammatory	Topical	[72]
Microspheres				
Zedoary oil	Sustained release and Higher bioavailability	Hepatoprotective	Oral	[73]
Quercetin	Significantly decreases the dose size	Anticancer	In vitro	[74]
Mentha spicata L. var. viridis oil (MVO)	Alleviate skin irritation and volatility and to avoid direct contact with the skin	Anti-fungal activity	Topical	[3]

and propylene glycol in the ratio of 3:24:38:38, respectively. Dispersion of the final composition of SNEDDSs formed spontaneously and converted in the form of nanoemulsion with a globule size of 64.80 ± 3.58 nm. Increased dispersion of the tested drug showed increased drug solubility, rapid absorption rate, and enhanced area under the curve, together with increased permeation and decreased efflux. All these properties led to the relative bioavailability of SNEDDS that was enhanced up to 200.22% in comparison with market tablets. Authors justified the potential of SNEDDS in phytomedicine and proposed these carriers as a potential candidate for an oral dosage form of cyclovirobuxine D like drugs with improved bioavailability [26]. Nanoemulsion is one of the most studied carriers for the improvement of bioavailability of the drugs. Nanoemulsions are the colloidal particulate system, i.e. composed of an aqueous phase and an organic phase using a suitable stabilizing surfactant that is thermodynamically metastable isotropic system. Due to the presence of surfactant-based interface, it shows reversible alteration in the structure of the biological membrane and hence facilitates absorption of drug inside the biological membrane. The most important advantage of this carrier is its ability to load both hydrophilic as well as hydrophobic drugs. Table 2 describes various types of nanoemulsion developed for various application. Despite high solubility in water, oral bioavailability of breviscapine is very low. This is due to the limited stability and permeation in the GIT. Its oral bioavailability has been attempted to improve by using nanoemulsion where this nanocarrier was used to facilitate breviscapine absorption following the oral route of administration. It also helped to increase the stability of the drug inside it. Higher drug encapsulation and uniform droplets size helped in finding these attributes. The improvement in bioavailability when observed in terms of relative bioavailability, it was observed to be enhanced up to 249.7%. This was due to the ability of nanoemulsion in encapsulating breviscapine in a manner that it did not expose to the gastric environment directly [27].

5 Nano-Phytomedicine for Improved Target Ability

Target may be defined as a cell or group of cells in the minority, identified to be in the need of treatment. Improved target ability is meant to specify a drug target and help them achieve highest concentration of drug for optimum therapeutic effect. These are some special form of drug carrier systems that carry actives/pharmacologically active agents/medicaments selectively at the desired site of action and avoid their distribution in other tissues and organs. They are vectors, which sequester, retain drug, and transport or deliver it into the vicinity of the target cells. Specific sites of actives can be a capillary bed of the site of action, specific cell type, or intracellular region. In order to achieve the goal of target ability, one has to make a carrier that is equipped with an agent that might intentionally recognize particular target only. For instances, carrier must be able to recognize tumor cells not normal cells. The ligands help in recognition and make carriers site specific when carrier lend drugs to approach the respective target. Some of the antibodies, polypeptides, endogenous hormones are specific examples of ligand systems. The target ability of the drugs is improved primarily in the treatment or prevention of diseases. Instability of the drugs in gastric environment, the low solubility of the drugs, lower GI absorption, high-plasma protein binding, biological instability, pharmacokinetic/pharmacodynamic short biological half-life, large volume of distribution, low specificity, low therapeutic index are some of the parameters that are considered while developing targeted drug delivery systems. The prime purpose of this approach is to minimize side effects associated with the distribution of the drug into other tissues. Targeted drug delivery systems should be biologically inert, non-immunogenic, physicochemically stable, should be able to restrict drug distribution to target cells/tissues/organs and should have a uniform capillary distribution. It should be controllable and should produce an optimum therapeutic dose of the drug. Carriers are designed to show minimal drug leakage during transit. It is also equally important whether polymers used for making carriers are biodegradable and readily eliminated from the body or not. These preparations should be easy in preparation or reasonably simple, reproductive, and cost-effective. These carriers are prepared by considering specific properties of target cells, nature of the markers, or transport carriers or vehicles, which convey drug to specific receptors with the help of ligands and physically modulated components. Cyclovirobuxine D (CVB-D) is a steroidal alkaloid, which is extracted from the Chinese medicinal herb Buxus microphylla. It is widely used to treat cerebrovascular diseases and other cardiovascular disorders. Very few preparations of CVB-D are investigated therefore, it was needed to improve its clinical application. One of the other drugs, i.e. thymoquinone, which is collected from the volatile oil of black cumin seeds (Nigella sativa) acts as a potential bioactive phytoconstituent in triple-negative breast cancer. As reported, it shows antineoplastic action on different molecular targets, including p53, p73 STST3, nuclear factor-κB, PPAR-γ, ROS, etc. in cancer [28]. Though many attempts have been made in the form of solid self-microemulsifying drug delivery systems, PEGylated forms, dispersible tablets, sustained release pellets, dripping pills, and phospholipid complexes only for oral administration and intravenous administration. CVB-D delivery to the brain was desired which was not achieved due to its restricted availability into the brain by the attempted route of administration. Only by making its intranasal drug delivery system, a working group of Wei, Liu et al., has been able to target CVB-D into the brain and that helped maintain its therapeutic level for the desired effect. Intranasal route helps drugs reach directly into the brain tissues or the cerebrospinal fluid (CSF) through the olfactory or trigeminal region of the nasal cavity, that bypasses BBB of the brain and avoid systemic drug side effects. The technology behind CVB-D targeting to the brain was angiopep-conjugated Polysorbate 80-coated liposomes. Liposomes were prepared to modify the surface of CVB-D. Synthesized DSPE-PEG2000-Angiopep-2 was used for the encapsulation of cyclovirobuxine D in Angiopep-conjugated Polysorbate 80-Coated Liposomes (T80-An2-CVB-D-Lps) prepared by thin film evaporation and an ultrasonic technique. It tackled the major challenge of blood–brain barrier (BBB) in the treatment of cerebrovascular diseases like cerebral ischemia-reperfusion injury, Parkinson's syndrome, and cerebral tumors. Angiopep-2-conjugated liposomes along with their intrinsic ability to cross the BBB bind to the low-density lipoprotein receptor-related protein-1 (LRP1) and hence proved to be a good prospect as a targeting carrier for brain delivery. Today the nasal route of administration drawn attention in it brain targeting capability [29].

Nano-phytomedicine has shown immense potential in targeting liver also [30]. Despite enormous efforts, liver disorders affect millions worldwide. Proper treatment of liver associated diseases is the need of the hour. Reasons for these diseases could be various types of liver injuries that include plants based, drugs, industrial toxins, water born, alcohol, and so on. However, several herbal medicinal preparations are used against disorders like cirrhosis, hepatitis, carcinomas, etc. Role of herbal source treatment options is now well proven against liver diseases. However, the optimum use of these treatments was ensured once the concept of targeted drug delivery systems was investigated first. Liver targeting is basically based on the concept of receptor ligand binding. Different types of the receptors that are available in the liver are found on the surface of hepatocytes, for instance, Kupffer cell, hepatic stellate cell, sinusoidal endothelial cells, etc. These are used to achieve an increased level of drug availability in the liver. Based on these receptor targets, various types of ligands such as galactosylated, lactobionic acid, asialofetuin are selected. These ligands are attached on the surface of various microparticles, liposomes, nanoparticles, and many more. The particulate approaches that have been used for liver targeting are liposomes, niosomes, nanoparticles, micelles, nanosuspensions, etc. [31]. Vinegar baked Radix Bupleuri (VBRB), traditional Chinese medicine is used to improve liver-targeting efficiency of the incorporated drug and has its own therapeutic effect. Ability of VBRB in improving target ability was investigated along with 10-hydroxycamptothecin (HCPT) after loading in polymeric micelles. VBRB effectively enhanced liver-targeting efficiency of HCPT loaded in polymeric micelles after oral co-administration. This was simple but effective strategy that reflects the potential use of meridian guided drug together with modern drug delivery system to achieve better active drug targeting [32]. A review summarized by Bartneck et al. in 2014 emphasizes the potential of nanomedicine in the treatment of liver inflammation and fibrosis by nanomedicine. Targeting of macrophages, hepatic stellate cells (HSC), hepatocytes, and liver sinusoidal endothelial cells (LSEC) are crucially involved in the complex sequence of events occurs during the initiation and maintenance of liver inflammation and fibrosis. Targeting of these cells is based on the capacity of these cells to ingest or engulf surrounding material with the help of RES systems through endocytosis and/or pinocytosis. The formulation scientists working in the area of nanomedicine pay special attention to the macrophages and liver-resident Kupffer cells. This is due to their highly efficient and unspecific uptake of most nanomaterials and due to their critical pathogenic behavior that helps recognize them by the macrophages and Kupffer cells expressed during inflammation and fibrogenesis. Macrophages might get activated in the presence of certain nanomaterials like peptide-modified gold nanorods (AuNRs). As far as targeting of macrophages in liver diseases is concerned, it can be done by using aiming mannose receptor. The collagen-producing cells, i.e. HSC is targeted by nanoconstructs that are able to recognize mannose 6-phosphate and insulin-like growth factor II, peroxisome proliferator activated receptor 1, platelet-derived growth factor (PDGF) receptor β, or integrins during fibrosis [33]. NanoCurcTM, i.e. polymeric nanoparticle based intra-peritoneal injection of curcumin has been developed to target HSC in CCl4-based fibrosis that shows induction of HSC apoptosis, and reduced ECM production. This approach helped efficient amelioration hepatic injury and fibrosis [34]. In another investigation, galactosyl receptor specific nanomedicine quercetin was developed to inject it subcutaneously for efficient binding on galactosyl receptor present over hepatocytes [35].

Bone drug delivery is restricted due to its unique histological features. As it is mainly composed of hydroxyapatite, one has to overcome this limitation while planning to deliver drugs to bone. Carriers that have an affinity towards hydroxyapatite are needed for targeted delivery. The reported carriers with bisphosphonates, alendronate (ALN), risedronate, and zoledronate are most often used for such purpose and are speedily dispersed to the bone surface following administration. P-C-P segment of bisphosphonate has a superior likeness for hydroxyapatite. In general, as drugs with a molecular weight of 30 kDa or less are susceptible to glomerular filtration by the kidney, those are either accumulated in the kidney or excreted in the urine. The utilization of macromolecular carriers consequently emerged to be a capable approach for efficient targeting to the bone [36].

A new bone-targeted drug delivery system of icaritin has been developed for the avoidance of steroid related osteonecrosis in small animals. This has been proved as an efficient strategy in the prevention of steroid-associated osteonecrosis. It works by inhibiting bone resorption and fat deposition and encourages bone development at osteonecrotic responsive skeletal targets. After identifying activity of novel phytomolecule (icaritin) on osteogenesis authors have developed its bone-targeted delivery system that can be used as anti-bone resorption and anti-adipogenesis effect. Asp8-liposome-icaritin significantly alleviated osteocytes apoptosis, down-regulated osteoclastogenesis, and up-regulated osteogenesis in steroids-treated rats; however, both liposome-icaritin and Asp8-liposome have not shown noteworthy effectiveness against increased SAON. Prevention of the bone resorption, suppression of adiposeness, and enhancement of bone-formation took place after the administration of Asp8-liposome-icaritin [37]. In addition to the liver, kidney, and tumors, hair follicles have also been targeted by nanomedicine. Anti-MRSA malleable liposomes facilitated hair follicle targeting and helped reach chloramphenicol for better treatment of hair follicles invaded with bacteria that are difficult to eradicate. Cutaneous and follicular delivery of chloramphenicol was ensured after loading it in to liposomes. Increased anti-MRSA potential of chloramphenicol was observed due to the presence of lipophilic dimyristoylphosphatidylcholine (DMPC), and deoxycholic acid (DA) in the developed liposomes, nanosized vesicles, and increased bilayer fluidity. By making these liposomes, twofolds higher activity was achieved in terms of inhibition of zone of growth inhibition. The deformable liposomes were reported to have good biocompatibility against keratinocytes and neutrophils and were safe to use on the skin [38]. These findings indicate the immense potential of nanomedicine in targeting of phytoconstituents to any level, i.e. cellular, tissue, or organ. These characteristics of nanomedicine are due to the versatile shape and structure of the nanocarriers, nano size range, surface charge, versatile polymer characteristics, and ultra-deformable integrity. Despite immense potential, targeting of phytoconstituents to the bone, tumor site, brain, liver, kidney, and to the specific cells needs additional efforts to place these carriers to the high scalable production process and to increase their clinical pertinence.

6 Nano-Phytomedicine for Improved Efficacy

Improved target ability leads to the increased efficacy. Most of the reports available on nanocarriers are for the improvement of efficacy against a particular disease. From modulation of physicochemical properties to modulation of biopharmaceutical, properties of the drug have been achieved by nanomedicines (Table 2). The pharmacokinetic profile of the phytoconstituents is majorly changed. From increasing absorption to modulating distribution, metabolism and excretion pattern of the drug are aimed while improved efficacy against diseases is required. Nanoemulsion based gel of tea tree oil (TTO) is developed to improve efficacy against the bacterial strains that are specific to acne. A novel TTO nanogel has been proved as a competent medium for the dermal TTO delivery with improved therapeutic effectiveness. This was due to the increased availability of TTO in different layers of the skin [39]. As microorganisms get more resistant to the antibiotics, the importance of the natural components with antimicrobial effect comes to the picture with great importance in avoiding such shortcomings. Matouskova et al. have checked the consequence of entrapment of herbal extracts in liposomes on antimicrobial efficacy and stability. Antimicrobial components entrapped in lipid or polysaccharide carriers have been found to exhibit good stability in aqueous conditions. Antimicrobial efficacy observed to be maintained in all types of particles. Liposomes with entrapped herbs and spice extracts showed significant efficacy against all assessed bacterial strains.

Particles loaded with both herbs and lysozyme exhibited higher activity against various kinds of bacteria, and high stability during long-term storage. The encapsulated plant extracts along with lysozyme proved as a composite antimicrobial formulation with a restricted release in the making of food and food supplements, pharmaceutical and cosmetic industries [40]. A review published by Hajialyani et al. highlights the application of natural product-based nanotechnological methods in improving wound healing and discussed the involved therapeutic targets. Nanotechnologies have been developed to modulate various pharmacological sites that are implicated in the wound healing properties of plant-based nanocomposites. Alleviating inflammatory cytokines, modulating inflammatory transduction cascades, alleviating oxidative factors, increasing antioxidative enzymes, and encouraging neovascularization and angiogenic pathways by improving the expression of vascular endothelial growth factor, fibroblast growth factor, and platelet-derived growth factor are some of them that have been investigated well [41]. In an example, simple nanoliposomes encapsulating lycium barbarum polysaccharides were developed to increase humoral and cellular immunity in small animal models. Targeted delivery and sustained release were the major attempts that have been made by preparing nanoliposomes. This approach improved immune responses. High drug loading, long-term stability of LBPL-OVA were linked with increased uptake of OVA by critical dendritic cells in the draining lymph nodes and they proficiently increased CD4+ and CD8+ T cell proliferation in vivo, further increasing antibody expression. These responses brought out noteworthy humoral and cellular immune responses, which were higher to that expressed alone by free antigens [42]. This application of nano-phytomedicine from wound healing to cancer treatment and immunity buster led these technologies to the next level and now being used in molecular imaging like MRI, CT scan, ultrasound, optical imaging, photoacoustic imaging, and positron emission tomography.

7 Nano-Phytomedicine for Improved Safety

Modification of physicochemical properties of the drugs portrays increased safety too. Investigations to improve safety highlight the potential nano-phytomedicine in safe drug delivery. Many of the investigations carried out recently prove it. Improved safety is another application that is achieved by utilizing nanomedicine approaches and these carriers are helping immensely to overcome such situations. Thymoquinone-rich fraction nanoemulsion (TQRFNE) was intended to improve the potential acute toxicity of thymoquinone. TQRFNE reduced toxicity associated with this drug when investigated in Sprague Dawley Rats. The parameters like common behavior, body weight, food and water consumption, relative organ weight, hematology, histopathology, and clinical biochemistry in the case of TQRFNE were observed to be unaltered in comparison to that of distilled water treated group and no mortality and symbol of toxicity were reported while the experiments commenced [43]. Conjugated-polymer-based ratiometric nanoprobe has been evaluated to be effective against in vivo herb-induced liver injury via MSOT imaging. This ratiometric nanoprobe was designed for in vivo imaging of hepatic injury with MSOT imaging that helps identify the accurate liver injury site. That liposomal nanoprobe consisted of a responsive dye (IX-2NH2) which could particularly react to NO and the diketopyrrolopyrrole-based conjugated polymer (DPP-TT) as the internal standard. This ratiometric optoacoustic recognition of herbal-medicine-induced liver injury with 3D information in the mouse model was recognized as a noninvasive way for the improvement of toxicity of the liver [44]. Various such types of techniques have been developed till date. Molecular imaging by nanomedicine itself has been identified as broad field of safety pharmacology. Techniques such as MRI, CT scan, ultrasound, optical and photoacoustic imaging, and positron emission tomography are currently being implemented to improve the safety of the patients.

8 Nano-Phytomedicine with Therapeutic Effects

Nanotherapeutics incorporating miRNA [45], bioactive lipids, and proteins work as extracellular envoy to arbitrate cell-cell communication in a way comparable to the exosomes secreted by mammalian cells. Chemically synthesized nanoparticles are applied to help reach low dose drugs (e.g., siRNAs, proteins, and peptides) to the exact target though these demonstrate two main restrictions. The components of the synthesized nanoparticles must be assessed for probable in vivo safety prior to clinical relevance and low fabrication scalability. However, nanoparticles formed from plant resources thought to surmount these shortcomings. Markedly, such nanoparticles are safe for oral purposes. These could work in interspecies communication and may work as natural therapeutics against various diseases. In addition, nanoparticles developed using plant based lipids are used to deliver drugs specifically at the desired site. Plant based edible nanoparticles could easily be scaled up for large-scale production as compared to that of the nanoparticles of synthetic origin.

Nanocarriers derived from juice of grapes, ginger, carrots, and tomatoes are extracted using an eco-friendly protocols. These nano-phytocomposites for the development of nanoparticles of therapeutic use are the substance of new approach in the epoch of nanomedicine [46]. Nanoparticles of herbal origin play a key role in interspecies communication, and function against inflammatory diseases and cancers. Recently, it has been observed that oral incorporation of the grapes exosome based nanocarriers to mice allegedly activated the cell proliferation in the intestinal epithelium, and was connected with the intestinal stem cell proliferation all through the intestine. In addition, grape exosome-like nanoparticles of oral use seemed to be resistant to degradation by saliva, gastric fluid, and proteolytic enzymes present in the intestinal tract. These current annotations recommended that edible-plant-based nanoparticles can be delivered to the intestine following the oral route, where they may be taken up by intestinal cells and exert tasks like intestinal regeneration. Edible nanoparticles able to target colonic tissues and having anti-inflammatory properties may symbolize novel natural and nontoxic delivery system that could simply be scaled up on large scale for the management of digestive tract discomfort such as intestinal bowel disease [47].

Lipids at present are the carriers of choice for efficient delivery of the drugs, however, synthesized liposomes and nanoparticles of lipidic nature offer an unintended property, for instances, cell stress, inflammasome activation, and programmed cell death. The utilization of safe "natural" nanovectors accumulated from natural nanoparticle-derived lipids are recommended as a reasonable optional approach for in vivo drug delivery. Ginger nanoparticles (Size about 180 nm, surface charge about −5 mV) showed the viability of produce "natural" nanovectors from ginger nanoparticles-derived lipids. The produced lipid nanoparticles had no cytotoxic effect on intestinal barrier function, signifying their utility in the effective delivery of the drug. Wang et al. demonstrated the utility of grapefruit-derived lipidic nanoparticles in delivering anti-cancerous drugs, phytoactive, DNA expression vectors, and proteins to cellular level. They also used grapefruit-derived nanovectors for co-delivery of actives along with folic acid. This approach has increased the concentration of the actives at the specific target such as the cells that expressed folate receptors. The nanovectors approach improved the inhibition of tumor growth in CT26- and SW620-cell-derived tumors of mice model. Grapefruit-derived nanovectors have demonstrated to be safe as compared to that of the synthetic lipid based nanoparticles. In addition, these did not show the drug availability in the placental matrix after intravenous injection into pregnant mice that prove their utility in drug delivery [48].

Eatable nanoparticles of the herbal source might have built-in beneficial actions against particular disorders. Investigators are presently looking to choose and distinguish explicit populations of nanoparticles of herbal source, which have such effectiveness and/or inspect their built-in specificity. It is expected that such studies would considerably add to the expansion of knowledge related to natural nanophytomedicines. Lipid nanoparticles of herbal source might perform as natural drug carriers that can deliver drug well on specific target. Therefore, eatable polymeric and lipid nanoparticles of herbal source with several profit (for example, nanoparticle size, low toxicity, good tissue-specific targeting, excellent environmental safety, and a significant potential for large-scale production) might provide as forthcoming beneficial delivery systems for the management of life-threatening and inflammatory disorders [46]. Metallic nanoparticles are used as a therapeutic agent and sometimes show additional or beneficial effect with other therapeutic agents. Some of the metallic nanoparticles that have been developed are Ag, Au, Fe, etc. Chemical process of synthesis of metallic nanoparticles is hazardous and produces toxicity after ingestion. Therefore, "green" synthesis processes in chemistry are becoming trendy nowadays. These protocols are much needed in lieu of environmental concerns. Silver is the most widely used nanomaterials with about 500 tons of manufacture of silver nanoparticles annually. Their production is thought to grow significantly in the forthcoming years. Together with its insightful task in the area of high sensitivity biomolecular recognition, catalysis, biosensors, and medicine, it has been recognized to have powerful inhibitory and bactericidal properties together with the anti-fungal, anti-inflammatory, and anti-angiogenesis actions. An array of methods are on hand for the synthesis of silver nanoparticles like ion sputtering, chemical reduction, sol gel, etc. [49]. Khan et al. summarized some novel reducing agent for the formation of silver nanoparticles. This reducing agent was derived from the plant extracts and termed it as green reductants for the synthesis of silver nanoparticles. This investigation was carried out in lieu of increasing demand for silver nanoparticles (Ag-NPs) by the pharmaceutical industries. These nanoparticles are used in the field of biomedicine owing to their antibacterial, anti-fungal, or anticancer properties. Chemical synthesis process involves the efficient production of the nanoparticles of desired size, morphology, and crystallinity; however, the requirement of hazardous chemicals in this chemical process calls for the exploration of some alternative way. Novel bio-inspired synthetic protocols have been investigated and these have played some promising role in minimizing harmful environmental effects [50, 51]; however, these biological protocols lack control over size, morphology, and crystallinity of the developed Ag-NPs. Their compilation briefs about chemical vs green protocols for Ag-NPs synthesis. The concept of green synthesis was based on the limitations associated with chemical and biological synthesis. Development of Ag-NPs of desired size, morphology, shape, crystallinity, and biological effect with minimal use of hazardous substances has been done using various plant extracts and phytomolecules and their chemical and biological properties, and their action on particle synthesis was portrayed. Some of the examples of plant extracts and phytomolecules that have been used in the green synthesis of Ag-NPs are extract of marigold flower, *Ziziphora tenuior*, *Abutilon indicum*, *Solanum tricobatum*, *Erythrina indica*, beet root, mangosteen, *Ocimum tenuiflorum*, *Spirogyra varians*, *Melia dubia*, olive, leaf extract of *Acalypha indica* and of *Sesuvium portulacastrum*. These have been reported with the synthesis of silver-NPs size range from 5 to 20 nm. The surface functionalization of silver-NPs by phytomolecules and their biomedical relevances are of utmost interest that are now being explored [52]. Due to the antimicrobial properties of Ag-NPs, these have been applied extensively in the health industry, medicine, textile coatings, food storage, dye reduction, wound dressing, antiseptic creams, and various environmental applications [53, 54]. Elemental silver and its compounds have been applied efficiently in various types of bacterial and fungal infections. *A. indica* extract mediated silver nanoparticles have been investigated as a possible antibacterial agent. Silver nanoparticles produced by green synthesis have been proved to display superior antibacterial action against *E. coli* and *S. aureus*. Therefore, these have immense potential in the treatment of various kinds of infectious diseases and disorders.

9 Conclusion

Nano-structured drug delivery systems show immense potential in delivering phytomedicine, i.e. meant for therapeutic purposes. This is due to their tremendous ability in the improvement of solubility, pharmacokinetic parameters such as absorption, distribution, metabolism and excretion, bioavailability, target ability, efficacy, and safety. In addition, they itself utilize various therapeutic benefits. Phytomedicines along with therapeutic nanocarriers have also been investigated for their dual effect. The development of phytomedicine based nanocarrier systems has been done to address the various unmet medical conditions. Nano-structured drug delivery systems developed till date are found to assist the due delivery of phytomedicines to the closest vicinity of diseased site because of their unique capability of modifying physicochemical properties of the drugs. If these proof of concepts are scalable for large scale translation, these have potential to get transferred from bench scale to the industrial level.

References

1. Rai VK, Mishra N, Yadav KS, Yadav NP (2018) Nanoemulsion as pharmaceutical carrier for dermal and transdermal drug delivery: formulation development, stability issues, basic considerations and applications. J Control Release 270:203–225. https://doi.org/10.1016/j.jconrel.2017.11.049
2. Mishra N, Yadav KS, Rai VK, Yadav NP (2017) Polysaccharide encrusted multilayered nano-colloidal system of andrographolide for improved hepatoprotection. AAPS Pharm Sci Tech 18 (2):381–392. https://doi.org/10.1208/s12249-016-0512-4
3. Mishra N, Rai VK, Yadav KS, Sinha P, Kanaujia A, Chanda D et al (2016) Encapsulation of mentha oil in chitosan polymer matrix alleviates skin irritation. AAPS Pharm Sci Tech 17 (2):482–492. https://doi.org/10.1208/s12249-015-0378-x
4. Sinha P, Srivastava N, Rai VK, Mishra R, Ajayakumar PV, Yadav NP (2019) A novel approach for dermal controlled release of salicylic acid for improved anti-inflammatory action: combination of hydrophilic-lipophilic balance and response surface methodology. J Drug Deliv Sci Technol 52:870–884. https://doi.org/10.1016/j.jddst.2019.06.007
5. Srivastava N, Patel DK, Rai VK, Pal A, Yadav NP (2018) Development of emulgel formulation for vaginal candidiasis: pharmaceutical characterization, in vitro and in vivo evaluation. J Drug Deliv Sci Technol 48:490–498. https://doi.org/10.1016/j.jddst.2018.10.013
6. Rai V, Sharma A, Upadhyay D, Sarma G, Kaur N, Gupta G et al (2020) Squalene integrated NLC based gel of tamoxifen citrate for efficient treatment of psoriasis: a preclinical investigation. J Drug Deliv Sci Technol 56:101568

7. Kumar B, Pandey M, Pottoo FH, Fayaz F, Sharma A, Sahoo PK (2020) Liposomes: novel drug delivery approach for targeting Parkinson's disease. Curr Pharm Des. https://doi.org/10.2174/1381612826666200128145124
8. Sharma S, Rabbani SA, Narang JK, Hyder Pottoo F, Ali J, Kumar S et al (2020) Role of rutin nanoemulsion in ameliorating oxidative stress: pharmacokinetic and pharmacodynamics studies. Chem Phys Lipids 228:104890. https://doi.org/10.1016/j.chemphyslip.2020.104890
9. Alam MS, Javed M, Pottoo F, Waziri A, Almalki F, Hasnain M et al (2019) QbD approached comparison of reaction mechanism in microwave synthesized gold nanoparticles and their superior catalytic role against hazardous nirto-dye. Appl Organomet Chem 33:e5071. https://doi.org/10.1002/aoc.5071
10. Javed MN, Alam MS, Waziri A, Pottoo FH, Yadav AK, Hasnain MS et al (2019) Chapter 12 - QbD applications for the development of nanopharmaceutical products. In: Beg S, Hasnain MS (eds) Pharmaceutical quality by design. Academic, New York, pp 229–253
11. Bhawana BRK, Buttar HS, Jain VK, Jain N (2011) Curcumin nanoparticles: preparation, characterization, and antimicrobial study. J Agric Food Chem 59(5):2056–2061. https://doi.org/10.1021/jf104402t
12. Ghosh P, Bag S, Singha Roy A, Subramani E, Chaudhury K, Dasgupta S (2016) Solubility enhancement of morin and epicatechin through encapsulation in an albumin based nanoparticulate system and their anticancer activity against the MDA-MB-468 breast cancer cell line. RSC Adv 6(103):101415–101429. https://doi.org/10.1039/C6RA20441D
13. Piazzini V, Monteforte E, Luceri C, Bigagli E, Bilia AR, Bergonzi MC (2017) Nanoemulsion for improving solubility and permeability of Vitex agnus-castus extract: formulation and in vitro evaluation using PAMPA and Caco-2 approaches. Drug Deliv 24(1):380–390. https://doi.org/10.1080/10717544.2016.1256002
14. Ansari MA, Chung I-M, Rajakumar G, Alzohairy MA, Alomary MN, Thiruvengadam M et al (2020) A current nanoparticles approaches in nose to brain drug delivery and anticancer therapy - a review. Curr Pharm Des. https://doi.org/10.2174/1381612826666200116153912
15. Ansari MA, Badrealam KF, Alam A, Tufail S, Khalique G, Equbal MJ et al (2020) Recent nano-based therapeutic intervention of bioactive sesquiterpenes: prospects in cancer therapeutics. Curr Pharm Des. https://doi.org/10.2174/1381612826666200116151522
16. Ahmad N, Ahmad R, Al-Qudaihi A, Alaseel S, Fita I, Khalid M et al (2019) Preparation of a novel curcumin nanoemulsion by ultrasonication and its comparative effects in wound healing and the treatment of inflammation. RSC Adv 9:20192–20206. https://doi.org/10.1039/C9RA03102B
17. Barkat MA, Pottoo FH, Singh SP, Ahmad FJ (2018) Therapeutic intervention of aloe gel containing nano-sized and micron-sized silver sulfadiazine gel on second-degree burn: a comparative study. Int J Low Extrem Wounds 17(3):176–183. https://doi.org/10.1177/1534734618791860
18. Barkat MA, Ahmad I, Ali R, Singh SP, Pottoo FH et al (2017) Nanosuspension-based aloe vera gel of silver sulfadiazine with improved wound healing activity. AAPS Pharm Sci Tech 18(8):3274–3285. https://doi.org/10.1208/s12249-017-0817-y
19. Lin H, Xie Q, Huang X, Ban J, Wang B, Wei X et al (2018) Increased skin permeation efficiency of imperatorin via charged ultradeformable lipid vesicles for transdermal delivery. Int J Nanomedicine 13:831–842. https://doi.org/10.2147/ijn.s150086
20. Su J, Sripanidkulchai K, Hu Y, Chaiittianan R, Sripanidkulchai B (2013) Increased in situ intestinal absorption of phytoestrogenic diarylheptanoids from Curcuma comosa in nanoemulsions. AAPS Pharm Sci Tech 14(3):1055–1062. https://doi.org/10.1208/s12249-013-9996-3
21. Chang LW, Hou ML, Tsai TH (2014) Silymarin in liposomes and ethosomes: pharmacokinetics and tissue distribution in free-moving rats by high-performance liquid chromatography-tandem mass spectrometry. J Agric Food Chem 62(48):11657–11665. https://doi.org/10.1021/jf504139g

22. Yang W, Yu XC, Chen XY, Zhang L, Lu CT, Zhao YZ (2012) Pharmacokinetics and tissue distribution profile of icariin propylene glycol-liposome intraperitoneal injection in mice. J Pharm Pharmacol 64(2):190–198. https://doi.org/10.1111/j.2042-7158.2011.01388.x
23. Li N, Feng L, Tan Y, Xiang Y, Zhang R, Yang M (2018) Preparation, characterization, pharmacokinetics and biodistribution of baicalin-loaded liposome on cerebral ischemia-reperfusion after i.v. administration in rats. Molecules 23(7):1747. https://doi.org/10.3390/molecules23071747
24. Luo LH, Zheng PJ, Nie H, Chen YC, Tong D, Chen J et al (2016) Pharmacokinetics and tissue distribution of docetaxel liposome mediated by a novel galactosylated cholesterol derivatives synthesized by lipase-catalyzed esterification in non-aqueous phase. Drug Deliv 23 (4):1282–1290. https://doi.org/10.3109/10717544.2014.980525
25. Di Costanzo A, Angelico R (2019) Formulation strategies for enhancing the bioavailability of silymarin: the state of the art. Molecules 24(11):2155. https://doi.org/10.3390/molecules24112155
26. Ke Z, Hou X, Jia XB (2016) Design and optimization of self-nanoemulsifying drug delivery systems for improved bioavailability of cyclovirobuxine D. Drug Des Devel Ther 10:2049–2060. https://doi.org/10.2147/dddt.s106356
27. Ma Y, Li H, Guan S (2015) Enhancement of the oral bioavailability of breviscapine by nanoemulsions drug delivery system. Drug Dev Ind Pharm 41(2):177–182. https://doi.org/10.3109/03639045.2014.947510
28. Barkat MA, Ahmad J, Khan MA, Beg S, Ahmad FJ (2018) Insights into the targeting potential of thymoquinone for therapeutic intervention against triple-negative breast cancer. Curr Drug Targets 19(1):70–80. https://doi.org/10.2174/1389450118666170612095959
29. Wei H, Liu T, Jiang N, Zhou K, Yang K, Ning W et al (2018) A novel delivery system of cyclovirobuxine D for brain targeting: angiopep-conjugated polysorbate 80-coated liposomes via intranasal administration. J Biomed Nanotechnol 14(7):1252–1262. https://doi.org/10.1166/jbn.2018.2581
30. Barkat MA, Rizwanullah M, Beg S, Pottoo FH, Siddiqui S et al (2019) Paclitaxel-loaded nanolipidic carriers with improved oral bioavailability and anticancer activity against human liver carcinoma. AAPS Pharm Sci Tech 20(2):87. https://doi.org/10.1208/s12249-019-1304-4
31. Rohilla R, Garg T, Goyal AK, Rath G (2016) Herbal and polymeric approaches for liver-targeting drug delivery: novel strategies and their significance. Drug Deliv 23(5):1645–1661. https://doi.org/10.3109/10717544.2014.945018
32. Wu H, Yu T, Tian Y, Wang Y, Zhao R, Mao S (2018) Enhanced liver-targeting via coadministration of 10-hydroxycamptothecin polymeric micelles with vinegar baked Radix Bupleuri. Phytomedicine 44:1–8. https://doi.org/10.1016/j.phymed.2018.04.022
33. Bartneck M, Warzecha KT, Tacke F (2014) Therapeutic targeting of liver inflammation and fibrosis by nanomedicine. Hepatobiliary Surg Nutr 3(6):364–376. https://doi.org/10.3978/j.issn.2304-3881.2014.11.02
34. Bisht S, Khan MA, Bekhit M, Bai H, Cornish T, Mizuma M et al (2011) A polymeric nanoparticle formulation of curcumin (NanoCurc™) ameliorates CCl4-induced hepatic injury and fibrosis through reduction of pro-inflammatory cytokines and stellate cell activation. Lab Investig 91(9):1383–1395. https://doi.org/10.1038/labinvest.2011.86
35. Mandal AK, Das S, Basu MK, Chakrabarti RN, Das N (2007) Hepatoprotective activity of liposomal flavonoid against arsenite-induced liver fibrosis. J Pharmacol Exp Ther 320 (3):994–1001. https://doi.org/10.1124/jpet.106.114215
36. Yamashita S, Katsumi H, Sakane T, Yamamoto A (2018) Bone-targeting dendrimer for the delivery of methotrexate and treatment of bone metastasis. J Drug Target 26(9):818–828. https://doi.org/10.1080/1061186X.2018.1434659
37. Chen S, Zheng L, Zhang J, Wu H, Wang N, Tong W et al (2018) A novel bone targeting delivery system carrying phytomolecule icaritin for prevention of steroid-associated osteonecrosis in rats. Bone 106:52–60. https://doi.org/10.1016/j.bone.2017.09.011

38. Hsu CY, Yang SC, Sung CT, Weng YH, Fang JY (2017) Anti-MRSA malleable liposomes carrying chloramphenicol for ameliorating hair follicle targeting. Int J Nanomedicine 12:8227–8238. https://doi.org/10.2147/ijn.s147226
39. Sinha P, Srivastava S, Mishra N, Singh DK, Luqman S, Chanda D et al (2016) Development, optimization, and characterization of a novel tea tree oil nanogel using response surface methodology. Drug Dev Ind Pharm 42(9):1434–1445. https://doi.org/10.3109/03639045.2016.1141931
40. Matouskova P, Marova I, Bokrova J, Benesova P (2016) Effect of encapsulation on antimicrobial activity of herbal extracts with lysozyme. Food Technol Biotechnol 54(3):304–316. https://doi.org/10.17113/ftb.54.03.16.4413
41. Hajialyani M, Tewari D, Sobarzo-Sanchez E, Nabavi SM, Farzaei MH, Abdollahi M (2018) Natural product-based nanomedicines for wound healing purposes: therapeutic targets and drug delivery systems. Int J Nanomedicine 13:5023–5043. https://doi.org/10.2147/ijn.s174072
42. Bo R, Sun Y, Zhou S, Ou N, Gu P, Liu Z et al (2017) Simple nanoliposomes encapsulating Lycium barbarum polysaccharides as adjuvants improve humoral and cellular immunity in mice. Int J Nanomedicine 12:6289–6301. https://doi.org/10.2147/ijn.s136820
43. Tubesha Z, Imam MU, Mahmud R, Ismail M (2013) Study on the potential toxicity of a thymoquinone-rich fraction nanoemulsion in Sprague Dawley rats. Molecules 18 (7):7460–7472. https://doi.org/10.3390/molecules18077460
44. Wu Y, Sun L, Zeng F, Wu S (2019) A conjugated-polymer-based ratiometric nanoprobe for evaluating in-vivo hepatotoxicity induced by herbal medicine via MSOT imaging. Photoacoustics 13:6–17. https://doi.org/10.1016/j.pacs.2018.11.002
45. Pottoo FH, Barkat MA, Ansari MA, Javed MN, Sajid Jamal QM et al (2019) Nanotechnological based miRNA intervention in the therapeutic management of neuroblastoma. Sem Cancer Biol. https://doi.org/10.1016/j.semcancer.2019.09.017
46. Zhang M, Viennois E, Xu C, Merlin D (2016) Plant derived edible nanoparticles as a new therapeutic approach against diseases. Tissue Barriers 4(2):e1134415. https://doi.org/10.1080/21688370.2015.1134415
47. Ju S, Mu J, Dokland T, Zhuang X, Wang Q, Jiang H et al (2013) Grape exosome-like nanoparticles induce intestinal stem cells and protect mice from DSS-induced colitis. Mol Ther 21(7):1345–1357. https://doi.org/10.1038/mt.2013.64
48. Wang Q, Zhuang X, Mu J, Deng ZB, Jiang H, Zhang L et al (2013) Delivery of therapeutic agents by nanoparticles made of grapefruit-derived lipids. Nat Commun 4:1867. https://doi.org/10.1038/ncomms2886
49. Ahmed S, Ahmad M, Swami BL, Ikram S (2016) Green synthesis of silver nanoparticles using Azadirachta indica aqueous leaf extract. J Radiat Res Appl Sci 9(1):1–7. https://doi.org/10.1016/j.jrras.2015.06.006
50. Mishra S, Sharma S, Javed MN, Pottoo FH, Barkat MA et al (2019) Bioinspired nanocomposites: applications in disease diagnosis and treatment. Pharm Nanotechnol 7 (3):206–219. https://doi.org/10.2174/2211738507666190425121509
51. Sharma S, Javed MN, Pottoo FH, Rabbani SA, Barkat MA et al (2019) Bioresponse inspired nanomaterials for targeted drug and gene delivery. Pharm Nanotechnol 7(3):220–233. https://doi.org/10.2174/2211738507666190429103814
52. Khan M, Shaik MR, Adil SF, Khan ST, Al-Warthan A, Siddiqui MRH et al (2018) Plant extracts as green reductants for the synthesis of silver nanoparticles: lessons from chemical synthesis. Dalton Trans 47(35):11988–12010. https://doi.org/10.1039/c8dt01152d
53. Ya Q, Cheng L, Li R, Liu G, Zhang Y, Tang X et al (2018) Potential antibacterial mechanism of silver nanoparticles and the optimization of orthopedic implants by advanced modification technologies. Int J Nanomedicine 13:3311–3327. https://doi.org/10.2147/IJN.S165125
54. Barkat MA, Beg S, Naim MJ, Pottoo FH, Singh SP et al (2018) Current progress in synthesis, characterization and applications of silver nanoparticles: precepts and prospects. Recent Pat Antiinfect Drug Discov 13(1):53–69. https://doi.org/10.2174/1574891X12666171006102833

55. El-Samaligy MS, Afifi NN, Mahmoud EA (2006) Increasing bioavailability of silymarin using a buccal liposomal delivery system: preparation and experimental design investigation. Int J Pharm 308(1-2):140–148. https://doi.org/10.1016/j.ijpharm.2005.11.006
56. Kesarwani K, Gupta R, Mukerjee A (2013) Bioavailability enhancers of herbal origin: an overview. Asian Pac J Trop Biomed 3(4):253–266. https://doi.org/10.1016/S2221-1691(13)60060-X
57. Ajazuddin SS (2010) Applications of novel drug delivery system for herbal formulations. Fitoterapia 81(7):680–689. https://doi.org/10.1016/j.fitote.2010.05.001
58. Mendonça EAM, Lira MCB, Rabello MM, Cavalcanti IMF, Galdino SL, Pitta IR et al (2012) Enhanced antiproliferative activity of the new anticancer candidate LPSF/AC04 in cyclodextrin inclusion complexes encapsulated into liposomes. AAPS Pharm Sci Tech 13(4):1355–1366. https://doi.org/10.1208/s12249-012-9853-9
59. Mosaddik A, Ravinayagam V, Elaanthikkal S, Fessi H, Badri W, Elaissari A (2018) Development and use of polymeric nanoparticles for the encapsulation and administration of plant extracts. In: Cechinel Filho V (ed) Natural products as source of molecules with therapeutic potential: research & development, challenges and perspectives. Springer, Cham, pp 391–463
60. Barupal AK, Gupta V, Ramteke S (2010) Preparation and characterization of ethosomes for topical delivery of aceclofenac. Indian J Pharm Sci 72(5):582–586. https://doi.org/10.4103/0250-474X.78524
61. Morsi NM, Aboelwafa AA, Dawoud MHS (2016) Improved bioavailability of timolol maleate via transdermal transfersomal gel: statistical optimization, characterization, and pharmacokinetic assessment. J Adv Res 7(5):691–701. https://doi.org/10.1016/j.jare.2016.07.003
62. Zhong H, Deng Y, Wang X, Yang B (2005) Multivesicular liposome formulation for the sustained delivery of breviscapine. Int J Pharm 301(1-2):15–24. https://doi.org/10.1016/j.ijpharm.2005.04.001
63. Mei Z, Chen H, Weng T, Yang Y, Yang X (2003) Solid lipid nanoparticle and microemulsion for topical delivery of triptolide. Eur J Pharm Biopharm 56(2):189–196
64. Yen FL, Wu TH, Lin LT, Cham TM, Lin CC (2008) Nanoparticles formulation of Cuscuta chinensis prevents acetaminophen-induced hepatotoxicity in rats. Food Chem Toxicol 46(5):1771–1777. https://doi.org/10.1016/j.fct.2008.01.021
65. Jain S, Jain S, Khare P, Gulbake A, Bansal D, Jain SK (2010) Design and development of solid lipid nanoparticles for topical delivery of an anti-fungal agent. Drug Deliv 17(6):443–451. https://doi.org/10.3109/10717544.2010.483252
66. Kothamasu P, Kanumur H, Ravur N, Maddu C, Parasuramrajam R, Thangavel S (2012) Nanocapsules: the weapons for novel drug delivery systems. Bioimpacts 2(2):71–81. https://doi.org/10.5681/bi.2012.011
67. Liu M, Li H, Luo G, Liu Q, Wang Y (2008) Pharmacokinetics and biodistribution of surface modification polymeric nanoparticles. Arch Pharm Res 31(4):547–554. https://doi.org/10.1007/s12272-001-1191-8
68. Min KH, Park K, Kim YS, Bae SM, Lee S, Jo HG et al (2008) Hydrophobically modified glycol chitosan nanoparticles-encapsulated camptothecin enhance the drug stability and tumor targeting in cancer therapy. J Control Release 127(3):208–218. https://doi.org/10.1016/j.jconrel.2008.01.013
69. Yanyu X, Yunmei S, Zhipeng C, Qineng P (2006) The preparation of silybin-phospholipid complex and the study on its pharmacokinetics in rats. Int J Pharm 307(1):77–82. https://doi.org/10.1016/j.ijpharm.2005.10.001
70. Maiti K, Mukherjee K, Murugan V, Saha BP, Mukherjee PK (2009) Exploring the effect of Hesperetin-HSPC complex--a novel drug delivery system on the in vitro release, therapeutic efficacy and pharmacokinetics. AAPS Pharm Sci Tech 10(3):943–950. https://doi.org/10.1208/s12249-009-9282-6
71. Zhao Y, Wang C, Chow AH, Ren K, Gong T, Zhang Z et al (2010) Self-nanoemulsifying drug delivery system (SNEDDS) for oral delivery of Zedoary essential oil: formulation and bioavailability studies. Int J Pharm 383(1-2):170–177. https://doi.org/10.1016/j.ijpharm.2009.08.035

72. Zheng Y, Hou SX, Chen T, Lu Y (2006) Preparation and characterization of transfersomes of three drugs in vitro. Chin J Chin Mater Med 31(9):728–731
73. You J, Cui FD, Han X, Wang YS, Yang L, Yu YW et al (2006) Study of the preparation of sustained-release microspheres containing zedoary turmeric oil by the emulsion-solvent-diffusion method and evaluation of the self-emulsification and bioavailability of the oil. Colloids Surf B Biointerfaces 48(1):35–41. https://doi.org/10.1016/j.colsurfb.2005.12.011
74. Chao P, Deshmukh M, Kutscher HL, Gao D, Rajan SS, Hu P et al (2010) Pulmonary targeting microparticulate camptothecin delivery system: anticancer evaluation in a rat orthotopic lung cancer model. Anti-Cancer Drugs 21(1):65–76. https://doi.org/10.1097/CAD.0b013e328332a322

Insights of Nanophytomedicines as a Combinatorial Therapy in Disease Diagnosis and Treatment

Akshay Kumar, Himanshi Walia, Faheem Hyder Pottoo, and Md. Noushad Javed

Abstract

Medicinal plants were the only cure until the late nineteenth century when German chemist Friedrich Wohler accidentally synthesized urea, the first organic synthesis in the history and a stepping stone for the era of synthetic compounds. In fact, the interest for natural life sources has expanded by 10% p.a. in North America, Europe, and Asia in the late decade. As per WHO, over 80% of the total population depends more regularly on conventional medications. Nature has been filled with the wellspring of therapeutic herbs. But the major issue with the phytomedicines is reduced bioavailability which leads to their low therapeutic index. Adaptation of nanomedicine techniques can hence be used to enhance the efficacy. Even better approach is the combinatorial therapy which would minimize the amount of adverse effects. While taking a two-drug combination, free form of one and the nanocarrier form of the other drug, or using separate nanocarriers for both the drugs, or co-encapsulating both the drugs in the same nanocarrier might be the possible options. Demonstration by various groups both in vitro and in vivo has proved the advantages of co-encapsulation when talking about synergistic drug combinations and using the same carrier. For example,

A. Kumar · H. Walia
Department of Pharmacology, Delhi Institute of Pharmaceutical Sciences and Research, Pushp Vihar, Delhi, India

F. H. Pottoo (✉)
Department of Pharmacology, College of Clinical Pharmacy, Imam Abdulrahman Bin Faisal University, Dammam, Saudi Arabia
e-mail: fhpottoo@iau.edu.sa

M. N. Javed
Department of Pharmaceutics, School of Pharmaceutical Sciences and Research, Jamia Hamdard, New Delhi, India

Department of Biotechnology, Jamia Millia Islamia, New Delhi, India
e-mail: noushad.sch@jamiahamdard.ac.in

S. Beg et al. (eds.), *Nanophytomedicine*,
https://doi.org/10.1007/978-981-15-4909-0_7

nanoluteolin exhibited a higher antiproliferative activity against the H292 cells having a lower IC_{50}-value when compared to free luteolin. The co-entrapment of two herbal drugs (glycyrrhizic acid and silibinin) to a nano-liposome(pegylated) exhibited synergistic antiproliferative effect on HCC cells. Silibinin had been shown to synergize the remedial efficacy of doxorubicin in advanced prostate cancer cells. Liposomal-resveratrol and liposomal-curcumin (2.5 mg/kg/each) in combination, reduced the incidence of prostate cancer mainly due to deprivation of *PTEN*, a tumor suppressor gene. The nanotechnology based phytomedicine therapy is now widely applied in the field of cancer with a tremendous amount of productive results.

Keywords

Nano-phytomedicine · Prostate cancer · Liver cancer · Breast cancer · Medicinal plants · Herbs

1 Introduction

To trace the history of phytomedicine or herbal drug therapy is to trace the ancient time itself. In early age, primitive man went in search of food and ate random plants. In case no harmful effects were observed he considered them edible and if some other effects were seen, they were considered as inedible [1]. According to the actions he used them in treating various symptoms and diseases. The results were passed on from one generation to the other and new knowledge was added in the same way. Medicinal plants were the only cure until the late nineteenth century when German chemist Friedrich Wohler was preparing ammonium cyanide from silver cyanide and ammonium chloride, and he accidentally synthesized urea. This was the first organic synthesis in the history and a stepping stone for the era of synthetic compound.

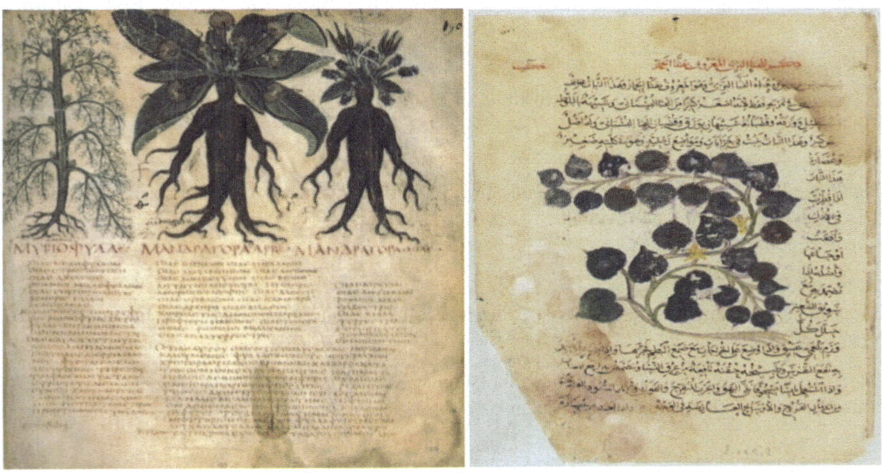

Pedanius Dioscorides, *De Materia Medica* (AD 65)

In excess of a tenth of plant species (500,000) have been utilized in pharmaceutical and corrective preparation. Be that as it may, the conveyance of plants containing dynamic phytoconstituents is not uniform all through the globe [2]. In fact, the interest in natural life sources has expanded by 10% p.a. in North America, Europe, and Asia in the late decade [3]. Numerous confirmations demonstrate that plants were developed as a medication for roughly 60,000 years ago [4, 5]. The ancient composition goes back to 5000 years in the Asian nations, China and Egypt and 2500 years in Central Asia and Greece [6]. The earliest written proof of the utilization of healthful plants for the purpose the medication was found on a Sumerian clay block almost 5000 years back [7]. Theophrastus, a Greek researcher, established the School of Medicinal Plants. At that point, Pedanius Dioscorides (a doctor and specialist) in the years 75–45 BC composed a reference book, called De Materia Medica, to depict 600 helpful therapeutic plants [8, 9]. After the accidental synthesis of urea, researchers started shifting their focus towards the synthetic drugs and following the next 100 years, phytomedicine was largely forgotten by the Western science. However, in the late 1980s there was a revival in the use of natural substances because of the derogatory effect of synthetic compounds. WHO characterized customary restorative plants as regular plant materials that were used at least or in the absence of industrial processing for the treatment of diseases at a local or regional scale [10].

As indicated by an information distributed by WHO, over 80% of the total population depends more regularly on conventional medications, for the most part plants, filling in as the fundamental wellspring of well-being care [11]. The details incorporate developing nations like India, in addition many progressed countries like China [12]. The significance of therapeutic plants and their items is progressively perceived and the open trust in their utilization is continually strengthened. In the present time, clinical, pharmaceutical, and concoction investigations of medications which are acquired from plants are the premise of numerous early medications, for example, Pilocarpine, Morphine, Aspirin, Digoxin, and Quinine, for purpose of treatment in a wide range of diseases [13]. Right now, it is evaluated that over half of the accessible medications are by one way or the another obtained from restorative plants. Phytotherapy is broadly being utilized over the world on a continually developing premise. Nature has been filled in as the wellspring of therapeutic herbs [14].

2 Introduction to Nanotechnology and Combinatorial Therapy

Incorporating the nanoformulation idea into herbal formulation research rendered huge number of benefits for phytomedicine, which includes better solubility and bioavailability, reduction in toxicity, site-specific action, increased pharmacological efficacy, stability improvement, protection against degradation, etc.

[15]. This involves implementation of nanoparticles presently under development and also a wide range of research that might involve the use of nano-robots to make alterations at the cellular level [16]. Hence, nanocarriers today are the topic of interest as possible alternatives for conventional therapeutic approaches. Nanotechnology aims in transforming the pharmaceutical field [17, 18]. A lot of medical field areas are being benefited constantly via this technology. Various types of nanovectors are being used for the delivery of phytodrugs such as micelles, cyclodextrins, nanocapsules, nanospheres, solid-lipid nanoparticles (SLN_S), cubosomes, limicubes, and liposomes [19]. One of the major issues with single drug therapy is drug resistance. However, it can be overcome by combining various drugs [20]. Combinatorial therapy tends to play a huge role when it comes to tackling complex diseased conditions. These benefits are mainly manifested by the arrest of more than one pathway, which play responsible role in different types of symptoms as well as diseased conditions [21]. A study suggests that this approach can be used in the case of tumors where targeting multiple pathways collectively eradicate a variety of clones that appear in a tumor [22]. Considering the huge investment cost being associated with the conventional experimentation methods that were used for selecting the appropriate drug combinations, it was necessary to shift to some better ways. This issue was solved by using computational approaches that limit the search area as well as reduce the cost [23, 24] (Fig. 1).

The major issue with phytomedicines is reduced bioavailability which leads to low therapeutic index. Adaptation of nanomedicine technique can hence be used to enhance the efficacy of the respective therapy. Further, this idea can be combined with the approach of combinatorial therapy which can be implemented for the diagnosis of a disease and its treatment. The basic idea behind combination therapy is to minimize the amount of side effects. However, this approach too might show a considerable level of toxicity. Therefore, selecting the drug candidates in a way that provides a non-overlapping toxicity profile should be the major area of focus. While taking a two-drug combination, free form of one and the nanocarrier form of the other drug (Free drug + Nano), or using separate nanocarriers for both the drugs (Nano + Nano), or co-encapsulating both the drugs in the same nanocarrier (co-encapsulation) might be the possible options. Demonstration by various groups both *in vitro* and *in vivo* has proved the advantages of co-encapsulation when talking about synergistic drug combinations and using the same carrier. The combination chemotherapy based selection rests on certain principles such as to use the drugs whose activities have been proven clinically as individual agents, minimal cross-drug resistance, choosing the candidates with different mechanisms of action, their ability to target different cell cycles, and last but not the least they should give rise to therapeutic synergy.

Nanotechnology-based medication conveyance is a noteworthy methodology that has changed the scene of pharmacotherapy. The nanomaterials (NMs) have been broadly applied for PK improvement and site-specific delivery and in this manner display a helpful conveyance apparatus to blend treatment [25, 26]. Nanomaterials help keep up a moderately indistinguishable in vivo destiny and the portion proportion of the combined medications. What's more, NMs can discharge the exemplified

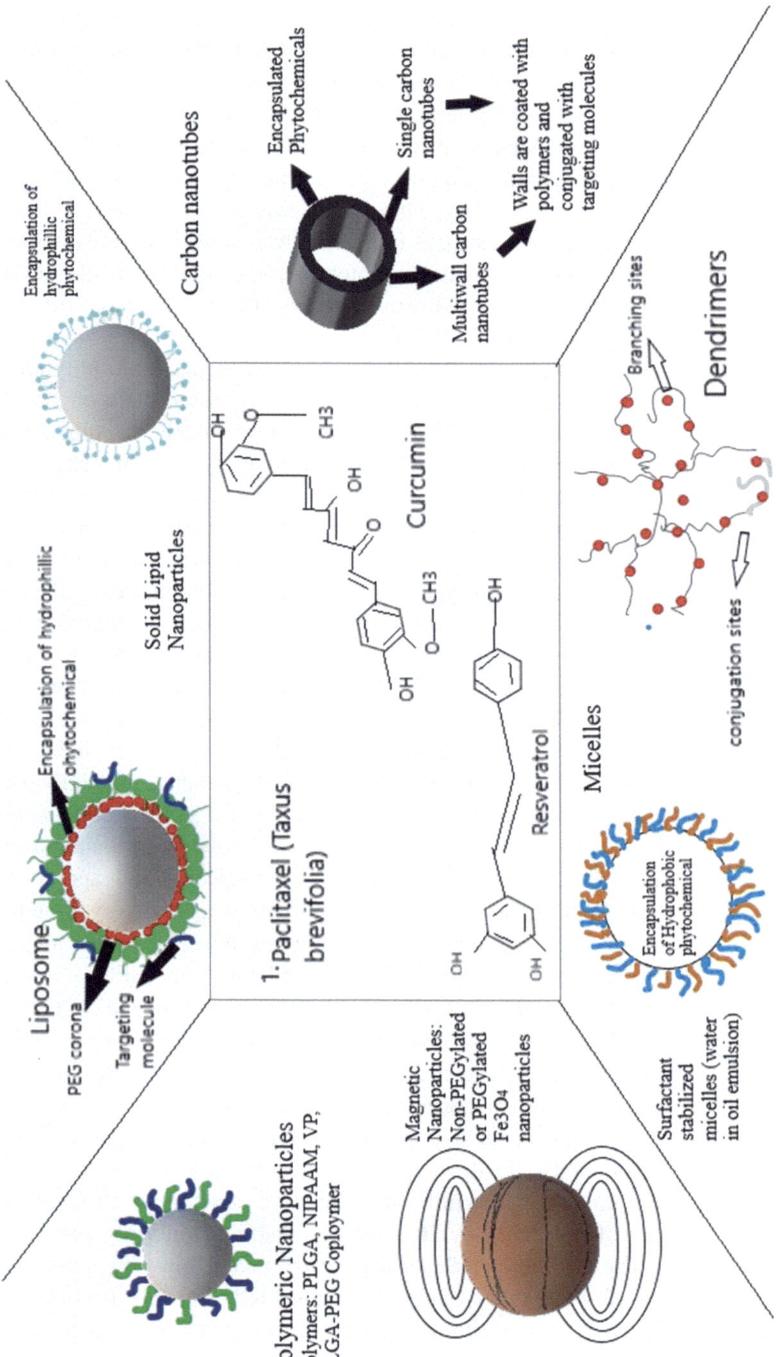

Fig. 1 Drug delivery systems for anticancer phytochemicals

medications in a controlled way. NMs are impervious to tranquilize efflux interceded by MDR transporters in view of the size-exclusion impact, and consequently the medications can hold a compelling intracellular concentration [27]. Complex infection, for example, malignancy, CVS, and neurological disease, for the most part includes multiple genes whose protein items assume significant roles in controlling distorted pathways and systems. Most medications are intended to specifically target explicit proteins as single medication medicines for the most part cannot breakdown the entire illness pathways and systems. This is the reason the customary "one illness, one gene, one medication" treatment frequently fizzles [28]. Various objective treatment, particularly nanophytomedicine combination can all the while focus on numerous segments of the sickness pathways and systems, in this way offering trust in treating such complex malady [29] (Table 1).

Wei et al., formulated honokiol micelles utilizing poly(3-caprolactone)-poly(ethylene glycol)-poly(3-caprolactone) copolymer (PCEC) and assessed them on A549 human lung adenocarcinoma cells. The outcomes demonstrated equivalent antiproliferative impact of free honokiol and honokiol-copolymeric micelles on A549 cells with similar application in the treatment of lung cancer [30]. Numerous phytodrug-stacked polymeric micelles for anticancer treatment are under scrutiny to improve sedate viability. A portion of these is in different periods of clinical preliminaries for, e.g., a square copolymer of PEG and polyglutamate (PGlu) conjugated with 7-ethyl-10-hydroxycamptothecin for breast cancer reemphasizing the current role for novel conveyance frameworks and extension for new helpful modalities for malignancy management. A study conducted by Merlina et al. involved the preparation of PLGA nanoparticles of ferulic acid via double emulsion technique using NCI-H460 cells as well as non-small-cell lung cancer cell lines for estimations. The results of the study showed an increase in the anticancer activity for ferulic acid nanoparticles versus a free form of ferulic acid. The ferulic acid nanoparticles induced cytotoxicity depicted an elevation in the altered mitochondrial transmembrane potential, DNA damage, reactive oxygen species and apoptotic changes with an overall betterment in the efficacy of the drug delivery system [31]. Gupta et al. arranged CUR nanoparticles utilizing silk fibroin polymer along with a bonded (covalently) mixture of silk fibroin and chitosan (SFCS) via a capillary-microdot method. Further evaluation was done on MDA-MB-453 and MCF-7 breast cancer cell lines. The results showed that silk fibroin based CUR nanoparticles showed an increased cellular uptake and a significant decrease in cell viability as compared to the SFCS nanoparticles in both the cell lines. Therefore, silk fibroin was proved to be a much accepted drug delivery vehicle in nano form and the use of it can be further extended to the development of nanoparticles of various other phytochemicals [32]. Another study utilized folic acid associated PEGylated liposomes of vincristine (FA-PEG-LS/VCR), for establishing multidrug resistant cancer chemo-therapy. The method included use of extrusion and thin-film hydration method. The evaluation of cytotoxicity was performed on KBv200 cells (multidrug resistant variant cell line). In vivo antitumor efficacy and in vitro MTT assay and

Table 1 Nanoformulated phytomedicine/s

S. No.	Phytochemical/s	Type of Nanoformulation	Effect	References
1.	Curcumin	Transferin conjugated SLNs of Curcumin	These newly formed nanoparticles target the Curcumin within the MCF-7 breast carcinoma tumor cells	[77]
2.	Curcumin	Curcumin-loaded γ-cyclodextrin Liposomal Nanoparticles	Cyclodextrin helps in raising the aqueous solvency of Curcumin	[78]
3.	Curcumin	Curcumin Nanoparticles	Tends to increase bioavailability (oral) of Curcumin	[79]
4.	Curcumin	CUR loaded into t hermoresponsive magnetic nanocarriers	Target Curcumin in cancer cell	[80]
5.	Doxorubicin + Resveratrol	PLGA Nanoparticle encapsulated DOX and RES	Downregulate expression of p-glycoprotein, nuclear factor expression leading to Mrp-I, BCRP, and induced apoptosis	[81]
6.	Doxorubicin + TRAIL	Co-delivery of TRAIL liposomes (TRAIL-LP) and DOX liposomes (DOX-LP)	Increase cellular apoptosis in MCF_&/ADR cells with a tumor inhibitory effect	[82]
7.	Curcumin + Resveratrol	Liposomes	Inhibition of cell growth and induction of apoptosis	[83]

studies along with apoptosis assessment studies and tumor growth inhibition studies were performed by TUNEL. As per the results, IC50 of the PEGylated folic acid conjugated VCR liposomes was found to be 23.99 nM when compared with 363.08 nM for PEG-LS/VCR 1.10 μM for free VCR. In vivo experiments depicted that the folic acid conjugation exclusively intensified the antitumor activity of PEGylated liposomes of VCR and also resulted in a higher apoptosis index as seen in the TUNEL assay (24.1 vs. 14.4% and 11.8% for free VCR) [33].

3 Cancer and Nano-Combinatorial Therapy

Many nanoparticle based therapies have been launched in the market in the past two decades for cancer management [34, 35]. Understanding the nanoparticle features (shape and size) has aided in phytonanoparticles development for various therapeutic applications. Cancer is a noxious disease that involves a lot of cellular abnormalities as well as genetic alterations [36, 37]. Such difficulties leads to mortality in patients. Multiple drug resistance is an issue that comes with the repeated use of single chemotherapeutic agent. There are various benefits associated with the nanotechnology-based drug systems especially when we take into account a combination of drugs, i.e better pharmacokinetic profile, and increased bioavailability of the drugs at the tumor site. The systemic toxicity to the normal tissue is also eliminated, by enhancing site-specific action [22].

3.1 Combinatorial Nanophytomedicine in the treatment of Lung Cancer

Lung carcinoma is a malignant uncontrolled cell growth in the tissues of the lung. During the progression of the disease, the tumor spreads to the other nearby tissues by the process of metastasis [38]. Two sorts of lung cancer are non-small cell lung carcinoma (NSCLC) and small cell lung carcinoma (SCLC). Lung malignancies start inside the cells covering the bronchi and segments of the respiratory organ like alveoli or bronchioles. About eightieth to eighty-fifth of respiratory organ malignant growths are NSCLC. The fundamental subtypes of NSCLC are squamous cell carcinoma, adenocarcinoma, and large cell carcinoma. About 12% of all the lung malignant growths are SCLC. This kind of lung malignancy will in general develop and spread quickly as compared to NSCLC. Since the malignant growth in this case develops rapidly, it will in general react well to chemotherapy and radiation treatment. Lamentably, for the vast majority, the malignant growth will return sooner or later [39, 40].

Getting exposed to arsenic either through inhalation or drinking water can result in lung carcinoma. "*Chelation therapy*" being the gold treatment for metal toxicity acts mainly by sequestering the intruding metal and formation of complexes that are easily excreted out from the body. However, clinical chelating agents are in a limited use because of their side effects. Also, they tend to exhibit issues with selectivity and specificity. Example being, "dimercaptosuccinic acid (DMSA)" which is commonly used for the treatment of lead poisoning in pediatric group and can also chelate arsenic and mercury [41]. This drug cannot cross the membrane of the cell and hence it is impossible for it to chelate metals from intracellular space [42]. All the issues have been overcome with the use of analogues of DMSA that are more lipophilic. The outcome of seven monoalkyl esters of DMSA which were different from each other on the lead mobilization in mice showed better reductions when treated with the monoalkylesters. The drug now targets the intracellular sites in the body that helps in an efficient elimination of the toxic metal ions from organelles inside the cell as well as the cytosol [43–45]. Curcumin is often an agent of choice because of the various properties it exhibits such as antioxidant effect, free radical-scavenging effect. It also shows metal-chelating properties that can be used in case of toxicities due to metal [46–48]. However, it has a low bioavailability due to poor absorption and rapid metabolism [49, 50].

Recent studies have revealed the potency of curcumin as (nanocurcumin) which is basically curcumin encapsulated chitosan nanoparticle in order to combat toxicity due to arsenic at much low doses when compared to free curcumin [51]. MiADMSA, a potential drug has been encapsulated in the hydrophobic core of co-polymeric nanoparticles of vinylpyrrolidone,*N*-isopropylacrylamide as well as acrylic acid (AA). Enhanced uptake due to this form has been reported when compared to the conventional form. According to the studies, it has been reported that all the treatment innovations which used nanoformulation (*NCur* + *NMiA*) exhibited the best arsenic removal profiles from tissues and blood. The better efficacy is a result of curcumin showing multiple effects including its antioxidant property as well as it being an adjuvant chelating agent. Luteolin was encapsulated by Majumdar et al. into polymeric NPs that comprised of PLA-PEG. H292 lung cancer cells were then used to evaluate them. The results indicated the nanoluteolin exhibited a higher antiproliferative activity by when against the H292 cells having a lower IC_{50}-value when compared to free luteolin [52].

Magnetic core-shell NPs were developed by Verma et al. for the aeros delivery of the drug, quercetin (QUR) by the process of nebulization and thereby evaluated the cytotoxicity in lung cancer cell line A549. The surface of MNPs was coated with PLGA in order to improve quercetin's dispersion, provision of stability and ensuring biocompatibility of the respective delivery system. In vitro and in vivo studies were carried out for evaluations. The results indicated a reduction in the count of viable A549 cells when talking about PLGA-MNPs loaded with Quercetin [53]. Co-encapsulated micelles of PTX and of β-lapachone using PEG-PLA copolymer were prepared by Zhang et al., which were then evaluated on A549 lung cancer cells. The study provided useful outcomes against lung cancer [54].

3.2 Combinatorial Nanophytomedicine in the treatment of Hepatocellular Carcinoma

Liver malignant growth also called hepatic cancer and essential hepatic malignancy is a malignancy that starts inside the liver. Cancer that has unfurled from somewhere else to the liver, called liver metastasis, has more events than that starts within the liver. Symptoms of the liver malady may grasp an irregularity beneath the skeletal structure, a lump or pain within the right facet below the skeletal structure, swelling of the stomach area, yellowish skin, weight reduction, and weakness [55].

The main cause is cirrhosis because of serum hepatitis, hepatitis C or alcohol abuse. Other causes incorporate aflatoxin, non-alcoholic fatty liver illness, and liver flukes. Blood tests, restorative imaging can be used in order to bolster the determination. Preventive endeavors incorporate inoculation against hepatitis B and treating those tainted with hepatitis B or C. Screening is suggested in those with chronic liver disease. Treatment choices include medical procedures, directed restorative consideration, and actinotherapy. In specific cases, removal treatment, embolization treatment, or liver transplantation might be used. Essential liver malignant growth is all-inclusive the 6th most successive disease (6%) and the second-driving reason for death from cancer (9%). In 2012, it involved 782,000 individuals and in 2015 it brought about 810,500 deaths. In 2015, 263,000 deaths from liver malignant growth were because of hepatitis B, 245,000 to liquor, and 167,000 to hepatitis

C. Regular anticancer medications cause different symptoms that outcome from non-specific toxicity and dissemination of the medication to typical cells [56]. Nano-liposomes are drug carriers with lipoid membranes used to improve drug delivery. Specifically, the anticancer potency of drugs increases with the decreasing size of the liposome [57]. Most investigations of Phyto-liposomes center around the liver-protectant flavonoid (silymarin-mixture) from the milk thorn [*Silybum marianum (L.)*], a restorative plant from Iran has been encountered. Silymarin's properties include immunomodulation, anti-fibrotic, membrane stabilizing, anti-oxidative, anti-inflammatory, and liver regeneration that play essential roles in experimental models of liver diseases [58]. The parts of silymarin are utilized for hepatoprotection while having poor fluid solvency and low bioavailability. In vivo investigations in regard to oral utilization of powdered parts of silymarin have indicated ng/ml raised concentration in blood plasma [59]. Different clinical and pharmacological impacts of silymarin have been accounted for, e.g., targeting cancer cell metastasis. Shockingly, Silymarin's poor dissolvability in water and oil results in pervasion through the intestinal epithelial film [60]. Silibinin is the most important antioxidant substance and the main biological active compound of silymarin [61]. Silibinin is an anticancer drug; its antitumor efficacy lies mainly in decreasing N-nitrosodiethylamine in liver cancer cells [62]. The advantage of the interaction between hydroxyl groups and the silibinin molecule is the rational identification of suitable sites for silibinin derivation while maintaining the biological activity of the resultant conjugates. Silibinin exerted a prohibitive effect on the encroachment and intrusion of advanced metastatic A549 cells although it showed a less pronounced effect considering adhesion. Silymarin mediated these effects by decreasing the expressions of matrix metalloproteases (MMP)-2 and u-PA and enhancing the expression of tissue inhibitor of MMP (TIMP)-2 [63]. Oral absorption of silibinin is limited and solubility of silibinin is poor in water and lipid media [64]. In previous studies, silymarin has been encapsulated to liposomes with a mixture of lecithin, cholesterol (CHOL), stearyl amine, and Tween 20 at a molar ratio of 9:1:1:0.5 as lipophilic phase having an 800 nm of particle size [65].

In addition to above, glycyrrhizic acid, a herbal drug obtained from *Glycyrrhiza glabra*(L.), was found to possess hepatoprotector, anti-viral, antitumor, and anti-inflammatory properties [66]. The antitumor efficacy of glycyrrhizic acid is due to the inhibition of MMPs and the protection of DNA in cancer cells. In previous studies such as Japan's drug of neo-minophagen c based glycyrrhizic acid was used for hepatic disease, Italy's drug of glycyrrhetinic acid phytosome was prepared based on liposomal glycyrrhizic acid, and Russia's drug of phosphogliv includes vegetative phospholipids with glycyrrhizic acid. All previous studies and products mainly focused on the preparation of liposome-based on one herbal drug, either glycyrrhizic acid or silymarin but in the study by Ochi et al., co-entrapment of two herbal drugs (glycyrrhizic acid and silibinin) to a Nano-liposome(pegylated) was effectuated and in vitro cytotoxicity of co-encapsulated Nano-liposomes was evaluated on a liver cancer cell line (HepG2) [67]. Nano-phyto liposomes (Nano-phytosome) are innovative formulations that improve the bioavailability of hydrophilic flavonoids. It has been demonstrated that polar phytoconstituents such as flavonoids, when mixed with phospholipids (e.g., phosphatidylcholine), generate a new drug delivery system called Nano-Phyto-liposome with a better absorption profile and more suitable lipid solubility which makes them competent to cross biological membranes [68]. Phyto-liposomes increased the specific activity of silymarin and had greater durability and stability than the free constituents. With respect to the antioxidant percentage and free radical cleaning properties in humans, liposomal silibinin effectively is absorbed into the target tissue of the liver. The results have suggested that the absorbency of liposomal silibinin is approximately seven times greater than that of free silibinin. The anti-hepatotoxic activity of liposomal silymarin is better than silymarin alone and can protect broiler chicks against the toxic effects of aflatoxin B1 [69]. Yanyu et al. prepared a silibinin–phospholipid complex, due to an impressive improvement, oral administration of phyto-liposomal silibinin increased bioavailability in rats and showed improved biological effect [70]. Different clinical studies have shown the bioavailability and pharmacological safety of silymarin. Experiments are now in progress to demonstrate the clinical influence of silymarin on different cancers. Silibinin is the most important antitumor substance of the silymarin compound and largely responsible for its anti-hepatotoxic activity [71].

In a study, co-encapsulation of two herbal antitumor drugs, silibinin and glycyrrhizic acid, to a pegylated nano-liposome was done. The *Encapsulation efficiency (EE)* for silibinin was greater than 24% and glycyrrhizic acid was over 68%. The two herbal drugs were co-entrapped to Nano-phyto liposome by hydrogen bond interactions between the phospholipids and the herbal drugs. These Nano-phyto liposomes could improve the delivery of phytomolecules [65].

In vitro cytotoxicity showed that the IC50 of co-encapsulated Nano-liposomes was lower than free silibinin (25% w/v) and glycyrrhizic acid (75% w/v). There was a considerable amount of difference in the cytotoxicity levels of the free form and that of the nanocarrier [67]. Silibinin was shown to seriously synergize the remedial efficacy of doxorubicin in advanced prostate cancer cells [72]. An in vitro study showed that Nano-liposome encapsulation of silibinin with glycyrrhizic acid increased the biological activity of free drugs, increased the stability of silibinin, and synergized the therapeutic effect of silibinin with glycyrrhizic acid on an HCC cell line.

3.3 Combinatorial Nanophytomedicine in the treatment of Breast Cancer

Functional foods are the things that have nutritive worth and are considered as an appropriate option for strong medications. Polyphenols are utilitarian nourishments that have a wide range of medical advantages [73]. Curcumin is a polyphenol that is orange-yellow in color and is hydrophobic. It is crystalline in nature [49].

Curcumin tends to upregulate various proteins like p16, Protein p53, p14ARF, and Cyclin-subordinate kinase (CDK) inhibitors. Curcumin invigorates apoptosis by the upregulation of proteins Bax, p53, Bad, Bim and decreasing the amount of B-cell lymphoma-additional enormous (Bcl-xL), B-cell lymphoma 2 (Bcl-2), mammalian target of rapamycin (mTOR) as well as Nuclear factor kappa-light-chain-enhancer of enacted B cells (NF-κB). Curcumin tends to regulate autophagy with the help of upregulation of lysosomal proteases, lipases, and phosphatases. By suppressing growth factors, curcumin tends to interfere with metastasis as well as angiogenesis. However most of the cancer cells show resistance to curcumin [74, 75]. The overexpression of multidrug resistance protein 1 (MDRP1), P-glycoprotein (P-gp), 2 (MDRP2), and breast cancer resistance protein (BCRP) might be the possible reasons for decreased effectiveness [76].

Moorthi et al. formulated Cu-Pi nanoparticles with PEG containing copolymer coating, as a dual nanoparticulate drug delivery system with overwhelming potential to overcome the limited oral bioavaliability and cancer cell targeting issues [73].

In the USA, breast cancer is the second most diagnosed cancer type in women, first being skin cancer. Survival rates for breast cancer have increased, and there has been a decline in the number of deaths mainly because of early detections, better approaches for treatment, and increased knowledge about the disease. Taking breast cancer into account, the early stages can be treated with hormone therapy that acts by decreasing the estrogen levels around the cancer cell. Aromatase inhibitors have been widely used for this purpose as aromatase is the enzyme that is involved in the production of estrogen [84]. Considering the MDR issues formulation of a novel PLN took place in which GG918 (elacridar) that is a chemosensitizer and doxorubicin were encapsulated. The results showed that the PLN co-encapsulated formulation exhibited a high efficiency as compared to the single agent formulation when talking about a MDR cell line of breast cancer [85]. Studies by Tanabe et al. illustrated that the combination of methotrexate and mitomycin C showed a good level of antitumor activity in breast cancer patients those who were initially treated with taxanes and anthracyclines [86]. Nanoparticles of human serum albumin 9HSA0 and noscapine have been prepared by Sebak et al. SK-BR-3 breast cancer cells were then treated using the combination by the use of pH coacervation method. The literature results showed that noscapine nanoparticles exhibit better efficacy as compared to the free noscapine when considering the reduction in the SK-BR-3 cell viability [87]. Co-encapsulating verapamil (VRP) and vincristine (VCR) into PLGA nanoparticles exhibited a significant antitumor profile when applied to MDR MCF-7/ADR xenograft models of human breast [88]. Many of polymer-integrated SLN (SLNPs) were formulated for the purpose of loading cytotoxic drugs such as camptothecin, doxorubicin, idarubicin 7-ethyl-10-hydroxy-20(S)-CPT, fluorodeoxyuridine (FudR), retinoic acid, ETP, and cholesteryl butyrate. These SLNPs were prepared by the use of those polymeric materials which were biodegradable, and non toxic [30]. A good combination of the drugs can lead to a coordinated drug release, with less toxicity and better efficacy. Acceptance of such systems could lead them into clinical studies and ultimately to the patient use. Further studies had demonstrated that berberine (BBR) attaches to DNA and radiosensitizes lung tumor cells via interaction with DNA repair related proteins [89, 90] and this resulted in the investigation that BBR enhances the anticancer activity of doxorubicin and rapamycin by making fine adjustment in their effect, interaction, and bioavailability [91, 92]. Researchers demonstrated that when BBR was given in combination with cisplatin it resulted in suppression of MCF-7 cell proliferation. This was confirmed by performing MTT assay on human breast cancer MCF-7 cell proliferation. A study involving BBR treatment for 48 h on MCF-7 cells showed that IC-50 value of BBR as 52.178 ± 1.593 μM, while IC-50 value of cisplatin as 49.541 ± 1.618 μM. The combination of 26 μM BBR to the cisplatin, reduced IC-50 value of cisplatin to 5.759 ± 0.76 μM. The result of study was that there was an increase in sensitivity of MCF-7 cells to cisplatin in a dose and time dependent manner. Berberine sensitizes MCF-7 cells to cisplatin through the caspase-3 dependent apoptotic pathway [93]. The results also confirmed that cisplatin actuated DNA breaks. It was also verified that when BBR given along with cisplatin, accelerates the DNA break initiated by cisplatin [93] (Table 2).

3.4 Combinatorial Nanophytomedicine in the treatment of Prostate Cancer

The prostate is around 3 cm long and weighs about 20 g. The location of prostate gland is in the male pelvis at the base of the penis [94]. It is the most common type of cancer diagnosed worldwide. It is the sixth leading cancer that is responsible for death in males [95]. Prostate-specific antigen (PSA) and prostate tissue biopsies are the main diagnostic parameters. According to the literature, one of the phenolic compounds, curcumin obtained from *Curcuma longa* has shown some positive antitumor effects [96]. Tumor bearing rodents have shown to better survival mainly due to tumor growth inhibition. Issue with bioavailability however persists when talking about the agents having natural origin. However as stated by recent studies, liposomal encapsulation might enhance the efficacy profile of curcumin [97, 98]. According to a study, two potential agents of natural origin namely resveratrol and curcumin were investigated in order to decrease the prevalence of prostate cancer. A limitation considering their bioavailability has however narrowed their clinical use. Hence, the study used curcumin and resveratrol that were liposome encapsulated individually and also in combination and were given in male B6C3F1/J mice. Study also included the estimation of chemopreventive measure of liposome encapsulated resveratrol and curcumin when given as a combination in prostate-specific PTEN knockout mice. Studies conducted in vitro reported that resveratrol and curcumin significantly inhibited cell growth plus induced apoptosis. Evidences from this study revealed the chemoprotective efficacy of the phytochemicals in combination form which can be used in the treatment of prostate cancer. Lipo-resveratrol and lipo-curcumin (2.5 mg/kg/each) in combination were mixed. Final products were stored at −20 °C. The study with clarity suggested that the combination form of phytochemicals could be used to decrease incidence of prostate cancer mainly due to deprivation of *PTEN*, a tumor suppressor gene [83]. Mice were administered with 250 μl of individual liposomal agents and in combination after an overnight fasting. Liposome alone was administered to control mice. The study showed that bioavailability of curcumin was enhanced when administered along

Table 2 Summary of challenges to breast cancer drug therapy and the way nanophytomedicine and combination can be used to tackle these challenges

Challenges to therapy	How Nanophytomedicine can help
Inefficient access to medications to metastatic locales, for example, cerebrum and bone	Nanomedicine improves cerebrum and bone infiltration
Undesirable pharmacokinetics such as quick clearance and short half-life	Use of techniques, for example, PEGlyation to broaden the dissemination time
The dose-restricting danger of the anticancer medications or the excipients, for instance, surfactants and natural co-solvents	Controlled medication discharge from nanocarrier; dissolvable, surfactant-free nanoformulation
Medication obstruction at cell level, for instance, expanded medication efflux	Nanoformulations may hinder drug efflux systems; co-conveyance of specialists that target drug obstruction mechanisms
Drug obstruction at tumor microenvironment level, for instance, lower pH, hypoxia, malignancy microenvironment crosstalk, etc.	Focusing on tumor microenvironment; utilization of boost responsive nanoformulations, for example, pH-responsive gadgets
Difficulty in eradicating cancer stem cells	Targeting cancer stem cell
Undesirable pharmaceutical properties of the medications, for instance, low watery dissolvability, poor in vivo stability	Many nanocarriers can accomplish medication solubilization and can ensure temperamental medications
Suboptimal dosing timetable and arrangement, particularly when mixes of different medications are involved	Careful streamlining of dosing calendar and succession; utilization of nanocarrier to co-convey different medications

with resveratrol. The overall results indicate increased level of both lipo-curcumin and lipo-resveratrol in the prostate tissue and serum. Another study was conducted in which resveratrol-loaded nanoparticles were used for the treatment of prostate cancer and was mainly based on the mixture of poly (D,L-lactic-co-glycolic acid)-poly (ethylene glycol) and poly(epsiloncaprolactone) [99]. Yet another study was conducted in which, Adenocarcinoma mouse prostate males were administered with resveratrol (625 mg resveratrol per kg AIN-76A diet) or a control diet (AIN-76A) initiating at an age of 5 weeks. Resveratrol majorly decreased the incidence of poorly separated prostatic adenocarcinoma. In the dorsolateral prostate, it showed inhibition in cell proliferation, an increase in estrogen receptor-beta, androgen receptor, and insulin-like growth factor-1 receptor, and a prominent decrease of phospho-extracellular regulating kinase 1 (phospho-ERK 1) and in insulin-like growth factor (IGF)-1. The downregulation of downstream effectors and reduced potent growth factor and cell proliferation, IGF-1, along with phospho-ERKs 1 and 2 provide with a basis for resveratrol based suppression of prostate cancer development [100] (Table 3).

Table 3 Nanoformulated phytomedicine in the treatment of prostate cancer

Therapeutic agents	Mechanism	Outcomes	References
Resveratrol	Cell proliferation was inhibited, increase in estrogen receptor-beta, androgen receptor, and insulin-like growth factor-1 receptors	Suppressed development of prostate cancer	[100]
Liposomal encapsulation of Resveratrol and Curcumin	Deprivation of PTEN	Inhibition of growth of cells	[83]

4 Conclusion

Hence, we conclude that phytomedicines which were at some point neglected by the people because of an increased emergence of modern allopathic medicines are now regaining their value. This is mainly because of comparatively high toxic profile associated with the modern drugs. In this chapter we discussed about nanophytomedicine which implies the incorporation of nanotechnology into natural agents. Introducing this technology into the conventional forms has provided a better pharmacokinetic and pharmacodynamic profile of the natural drugs mainly by enhancing bioavailability, reducing toxicity, and a better treatment efficacy. In next section we talked about combinatorial therapy which signified the use of two or more agents in order to have synergy because of multiple mechanisms of action of the respective drugs. Further, combining this approach to the nanophytomedicine has allowed us to use various phytoconstituents in nano form in order to provide the best results which can be used for the treatment of various diseases. This technology is now widely being adopted in the field of cancer chemotherapy of lungs, liver, breast, and prostate, in order to achieve significant amount of promising outcomes and productive results. Therefore, considering all such benefits this approach definitely has a long way to go for the betterment of health.

References

1. Alam MS, Garg A, Pottoo FH, Saifullah MK, Tareq AI, Manzoor O, Mohsin M, Javed MN (2017) Gum ghatti mediated, one pot green synthesis of optimized gold nanoparticles: investigation of process-variables impact using Box-Behnken based statistical design. Int J Biol Macromol 104(Pt A):758–767
2. Huang H (2011) Plant diversity and conservation in China: planning a strategic bioresource for a sustainable future. Bot J Linn Soc 166(3):282–300
3. Singh SV (2008) Current and future status of herbal medicines. Veterinary World. Available from http://agris.fao.org/agris-search/search.do?recordID=DJ2012059754
4. Solecki RS (1975) Shanidar IV, a Neanderthal flower burial in Northern Iraq. Science 190 (4217):880–881

5. Alam MS, Javed MN, Pottoo FH, Waziri A, Almalki FA, Hasnain MS, Garg A, Saifullah MK (2019) QbD approached comparison of reaction mechanism in microwave synthesized gold nanoparticles and their superior catalytic role against hazardous nirto-dye. Appl Organomet Chem 33(9):e5071
6. Ang-Lee MK, Moss J, Yuan CS (2001) Herbal medicines and perioperative care. JAMA 286 (2):208–216
7. Qiu J (2007) A culture in the balance. Nature 448(7150):126–128
8. Schipmann U, Leaman DJ, Cunningham AB, Walter S (2005) Impact of cultivation and collection on the conservation of medicinal plants: global trends and issues. Acta Hortic 676:31–44
9. Lindberg Madsen H, Bertelsen G (1995) Spices as antioxidants. Trends Food Sci Technol 6 (8):271–277
10. Tilburt JC, Kaptchuk TJ (2008) Herbal medicine research and global health: an ethical analysis. Bull World Health Organ 86(8):594–599
11. Farnsworth NR, Akerele O, Bingel AS, Soejarto DD, Guo Z (1985) Medicinal plants in therapy. Bull World Health Organ 63(6):965–981
12. Ganesan A (2008) The impact of natural products upon modern drug discovery. Curr Opin Chem Biol 12(3):306–317
13. Hasnain MS, Javed N, Alam MS, Rishishwar P, Rishishwar S, Ali S, Nayak AK, Beg S (2019) Purple heart plant leaves extract-mediated silver nanoparticle synthesis: optimization by Box-Behnken design. Mater Sci Eng 99:1105–1114
14. Fabricant DS, Farnsworth NR (2001) The value of plants used in traditional medicine for drug discovery. Environ Health Perspect 109(Suppl 1):69–75
15. Sharma S, Rabbani SA, Narang JK, Hyder Pottoo F, Ali J, Kumar S, Baboota S (2020) Role of Rutin nanoemulsion in ameliorating oxidative stress: pharmacokinetic and pharmacodynamics studies. Chem Phys Lipids 228:104890
16. Javed MN, Kohli K, Amin S (2018) Risk assessment integrated QbD approach for development of optimized bicontinuous mucoadhesive limicubes for oral delivery of rosuvastatin. AAPS Pharm Sci Tech 19(3):1377–1391
17. Mishra S, Sharma S, Javed MN, Pottoo FH, Abul Barkat M, Null H, Amir M, Sarfaroz M (2019) Bioinspired nanocomposites: applications in disease diagnosis and treatment. Pharm Nanotechnol 7(3):206–219
18. Sharma S, Noushad Javed M, Pottoo F, Arman Rabbani S, Barkat M, Ab H, Sarafroz D, Amir M (2019) Bioresponse inspired nanomaterials for targeted drug and gene delivery. Pharm Nanotechnol 7(3):220–233
19. Kumar B, Pandey M, Pottoo FH, Fayaz F, Sharma A, Sahoo PK (2020) Liposomes: novel drug delivery approach for targeting Parkinson's disease. Curr Pharm Des. https://doi.org/10.2174/1381612826666200128145124
20. Pottoo FH, Tabassum N, Javed MN, Nigar S, Sharma S, Barkat MA, Null H, Alam MS, Ansari MA, Barreto GE, Ashraf GM (2020) Raloxifene potentiates the effect of fluoxetine against maximal electroshock induced seizures in mice. Eur J Pharm Sci 146:105261
21. Pottoo FH, Tabassum N, Javed MN, Nigar S, Rasheed R, Khan A, Barkat MA, Alam MS, Maqbool A, Ansari MA, Barreto GE, Ashraf GM (2018) The synergistic effect of raloxifene, fluoxetine, and bromocriptine protects against pilocarpine-induced status epilepticus and temporal lobe epilepsy. Mol Neurobiol 56(2):1233–1247
22. Pottoo FH, Barkat A, Harshita, Ansari MA, Javed N, Sajid Jamal QM, Kamal MA (2019) Nanotechnological based miRNA intervention in the therapeutic management of neuroblastoma. Sem Cancer Biol. https://doi.org/10.1016/j.semcancer.2019.09.017
23. Nawaz F, Alam O, Perwez A, Rizvi MA, Naim MJ, Siddiqui N, Pottoo FH, Jha M (2020) 3'-(4-(Benzyloxy)phenyl)-1'-phenyl-5-(heteroaryl/aryl)-3,4-dihydro-1'H,2H-[3,4'-bipyrazole]-2-carboxamides as EGFR kinase inhibitors: synthesis, anticancer evaluation, and molecular docking studies. Arch Pharm 353:e1900262

24. El Rayes SM, Aboelmagd A, Gomaa MS, Ali IAI, Fathalla W, Pottoo FH, Khan FA (2019) Convenient synthesis and anticancer activity of methyl 2-[3-(3-phenyl-quinoxalin-2-ylsulfanyl)propanamido]alkanoates and N-alkyl 3-((3-phenyl-quinoxalin-2-yl)sulfanyl) propanamides. ACS Omega 4(20):18555–18566
25. Gao Z, Zhang L, Sun Y (2012) Nanotechnology applied to overcome tumor drug resistance. J Control Release 162(1):45–55
26. Ahmad N, Ahmad R, Alam MA, Ahmad FJ, Amir M, Pottoo FH, Sarafroz M, Jafar M, Umar K (2019) Daunorubicin oral bioavailability enhancement by surface coated natural biodegradable macromolecule chitosan based polymeric nanoparticles. Int J Biol Macromol 128:825–838
27. Liu J, Zhao Y, Guo Q, Wang Z, Wang H, Yang Y, Huang Y (2012) TAT-modified nanosilver for combating multidrug-resistant cancer. Biomaterials 33(26):6155–6161
28. Hopkins AL (2008) Network pharmacology: the next paradigm in drug discovery. Nat Chem Biol 4(11):682–690
29. Jia J, Zhu F, Ma X, Cao Z, Cao ZW, Li Y, Li YX, Chen YZ (2009) Mechanisms of drug combinations: interaction and network perspectives. Nat Rev Drug Discov 8(2):111–128
30. Wei T, Chen C, Liu J, Liu C, Posocco P, Liu X, Cheng Q, Huo S, Liang Z, Fermeglia M, Pricl S, Liang X-J, Rocchi P, Peng L (2015) Anticancer drug nanomicelles formed by self-assembling amphiphilic dendrimer to combat cancer drug resistance. PNAS 112 (10):2978–2983
31. Merlin JPJ, Rajendra Prasad N, Shibli SMA, Sebeela M (2012) Ferulic acid loaded Poly-d,l-lactide-co-glycolide nanoparticles: Systematic study of particle size, drug encapsulation efficiency and anticancer effect in non-small cell lung carcinoma cell line in vitro. Biomed Prev Nutr 2(1):69–76
32. Gupta V, Aseh A, Ríos CN, Aggarwal BB, Mathur AB (2009) Fabrication and characterization of silk fibroin-derived curcumin nanoparticles for cancer therapy. Int J Nanomedicine 4:115–122
33. Wang C, Feng L, Yang X, Wang F, Lu W (2013) Folic acid-conjugated liposomal vincristine for multidrug resistant cancer therapy. Asian J Pharm Sci 8(2):118–127
34. Barkat MA, Beg S, Pottoo FH, Ahmad FJ (2019) Nanopaclitaxel therapy: an evidence based review on the battle for next-generation formulation challenges. Nanomedicine 14 (10):1323–1341
35. Barkat MA, Rizwanullah M, Beg S, Pottoo FH, Siddiqui S, Ahmad FJ (2019) Paclitaxel-loaded nanolipidic carriers with improved oral bioavailability and anticancer activity against human liver carcinoma. AAPS Pharm Sci Tech 20(2):87
36. Ansari MA, Chung I-M, Rajakumar G, Alzohairy MA, Alomary MN, Thiruvengadam M, Pottoo FH, Ahmad N (2020) A current nanoparticles approaches in nose to brain drug delivery and anticancer therapy - a review. Curr Pharm Des. https://doi.org/10.2174/1381612826666200116153912
37. Ansari MA, Badrealam KF, Alam A, Tufail S, Khalique G, Equbal MJ, Alzohairy MA, Almatroudi A, Alomary MN, Pottoo FH (2020) Recent Nano-based therapeutic intervention of bioactive sesquiterpenes: prospects in cancer therapeutics. Curr Pharm Des. https://doi.org/10.2174/1381612826666200116151522
38. Lundin A, Driscoll B (2013) Lung cancer stem cells: progress and prospects. Cancer Lett 338 (1):89–93
39. Inamura K (2017) Lung cancer: understanding its molecular pathology and the 2015 WHO classification. Front Oncol 7:5581350
40. Petersen I (2011) The morphological and molecular diagnosis of lung cancer. Dtsch Arztebl Int 108(31–32):525–531
41. Flora SJ (1999) Arsenic-induced oxidative stress and its reversibility following combined administration of N-acetylcysteine and meso 2,3-dimercaptosuccinic acid in rats. Clin Exp Pharmacol Physiol 26(11):865–869

42. Aposhian HV, Maiorino RM, Rivera M, Bruce DC, Dart RC, Hurlbut KM, Levine DJ, Zheng W, Fernando Q, Carter D (1992) Human studies with the chelating agents, DMPS and DMSA. J Toxicol Clin Toxicol 30(4):505–528
43. Mehta A, Pant SC, Flora SJS (2006) Monoisoamyl dimercaptosuccinic acid induced changes in pregnant female rats during late gestation and lactation. Reprod Toxicol 21(1):94–103
44. Mehta A, Kannan GM, Dube SN, Pant BP, Pant SC, Flora SJS (2002) Haematological, hepatic and renal alterations after repeated oral or intraperitoneal administration of monoisoamyl DMSA. I. Changes in male rats. J Appl Toxicol 22(6):359–369
45. Modi M, Pathak U, Kalia K, Flora SJS (2005) Arsenic antagonism studies with monoisoamyl DMSA and zinc in male mice. Environ Toxicol Pharmacol 19(1):131–138
46. Flora SJS, Shrivastava R, Mittal M (2013) Chemistry and pharmacological properties of some natural and synthetic antioxidants for heavy metal toxicity. Curr Med Chem 20 (36):4540–4574
47. Ak T, Gülçin I (2008) Antioxidant and radical scavenging properties of curcumin. Chem Biol Interact 174(1):27–37
48. Shukla PK, Khanna VK, Khan MY, Srimal RC (2003) Protective effect of curcumin against lead neurotoxicity in rat. Hum Exp Toxicol 22(12):653–658
49. Ahmad N, Ahmad R, Al-Qudaihi A, Alaseel SE, Fita IZ, Khalid MS, Pottoo FH (2019) Preparation of a novel curcumin nanoemulsion by ultrasonication and its comparative effects in wound healing and the treatment of inflammation. RSC Adv 9(35):20192–20206
50. Ahmad N, Ahmad R, Al-Qudaihi A, Alaseel SE, Fita IZ, Khalid MS, Pottoo FH, Bolla SR (2019) A novel self-nanoemulsifying drug delivery system for curcumin used in the treatment of wound healing and inflammation. Biotechnology 9(10):360
51. Wahlström B, Blennow G (1978) A study on the fate of curcumin in the rat. Acta Pharmacol Toxicol 43(2):86–92
52. Majumdar D, Jung K-H, Zhang H, Nannapaneni S, Wang X, Amin ARMR, Chen Z, Chen ZG, Shin DM (2014) Luteolin nanoparticle in chemoprevention: in vitro and in vivo anticancer activity. Cancer Prev Res 7(1):65–73
53. Verma NK, Crosbie-Staunton K, Satti A, Gallagher S, Ryan KB, Doody T, McAtamney C, MacLoughlin R, Galvin P, Burke CS, Volkov Y, Gun'ko YK (2013) Magnetic core-shell nanoparticles for drug delivery by nebulization. J Nanobiotechnol 11(1):1
54. Zhang L, Chen Z, Yang K, Liu C, Gao J, Qian F (2015) β-lapachone and paclitaxel combination micelles with improved drug encapsulation and therapeutic synergy as novel nanotherapeutics for NQO1-targeted cancer therapy. Mol Pharm 12(11):3999–4010
55. Balogh J, Victor D, Asham EH, Burroughs SG, Boktour M, Saharia A, Li X, Ghobrial RM, Monsour HP (2016) Hepatocellular carcinoma: a review. J Hepatocell Carcinoma 3:41–53
56. Cheng WWK, Allen TM (2008) Targeted delivery of anti-CD19 liposomal doxorubicin in B-cell lymphoma: a comparison of whole monoclonal antibody, Fab' fragments and single chain Fv. J Control Release 126(1):50–58
57. Bozzuto G, Molinari A (2015) Liposomes as nanomedical devices. Int J Nanomedicine 10:975–999
58. Karimi G, Vahabzadeh M, Lari P, Rashedinia M, Moshiri M (2011) "Silymarin", a promising pharmacological agent for treatment of diseases. Iran J Basic Med Sci 14(4):308–317
59. Schandalik R, Gatti G, Perucca E (1992) Pharmacokinetics of silybin in bile following administration of silipide and silymarin in cholecystectomy patients. Arzneimittelforschung 42(7):964–968
60. Takke A, Shende P (2019) Nanotherapeutic silibinin: an insight of phytomedicine in healthcare reformation. Nanomedicine 21:102057
61. Bijak M (2017) Silybin, a major bioactive component of milk thistle (Silybum marianum L. Gaernt.)—chemistry, bioavailability, and metabolism. Molecules 22(11):6150307
62. Ramakrishnan G, Raghavendran HRB, Vinodhkumar R, Devaki T (2006) Suppression of N-nitrosodiethylamine induced hepatocarcinogenesis by silymarin in rats. Chem Biol Interact 161(2):104–114

63. Chu S-C, Chiou H-L, Chen P-N, Yang S-F, Hsieh Y-S (2004) Silibinin inhibits the invasion of human lung cancer cells via decreased productions of urokinase-plasminogen activator and matrix metalloproteinase-2. Mol Carcinog 40(3):143–149
64. El-Samaligy MS, Afifi NN, Mahmoud EA (2006) Increasing bioavailability of silymarin using a buccal liposomal delivery system: preparation and experimental design investigation. Int J Pharm 308(1–2):140–148
65. El-Samaligy MS, Afifi NN, Mahmoud EA (2006) Evaluation of hybrid liposomes-encapsulated silymarin regarding physical stability and in vivo performance. Int J Pharm 319(1–2):121–129
66. Li J, Cao H, Liu P, Cheng G, Sun M (2014) Glycyrrhizic acid in the treatment of liver diseases: literature review. Biomed Res Int 2014:4052927
67. Ochi MM, Amoabediny G, Rezayat SM, Akbarzadeh A, Ebrahimi B (2016) In vitro co-delivery evaluation of novel pegylated nano-liposomal herbal drugs of silibinin and glycyrrhizic acid (nano-phytosome) to hepatocellular carcinoma cells. Cell J 18(2):135–148
68. Post-White J, Ladas EJ, Kelly KM (2007) Advances in the use of milk thistle (Silybum marianum). Integr Cancer Ther 6(2):104–109
69. Amiridumari H, Sarir H, Afzali N, FaniMakki O (2013) Effects of milk thistle seed against aflatoxin B1 in broiler model. J Res Med Sci 18(9):786–790
70. Yanyu X, Yunmei S, Zhipeng C, Qineng P (2006) The preparation of silybin-phospholipid complex and the study on its pharmacokinetics in rats. Int J Pharm 307(1):77–82
71. Federico A, Dallio M, Loguercio C (2017) Silymarin/Silybin and chronic liver disease: a marriage of many years. Molecules 22(2):6155865
72. Tyagi AK, Singh RP, Agarwal C, Chan DCF, Agarwal R (2002) Silibinin strongly synergizes human prostate carcinoma DU145 cells to doxorubicin-induced growth inhibition, G2-M arrest, and apoptosis. Clin Cancer Res 8(11):3512–3519
73. Moorthi C, Krishnan K, Manavalan R, Kathiresan K (2012) Preparation and characterization of curcumin-piperine dual drug loaded nanoparticles. Asian Pac J Trop Biomed 2 (11):841–848
74. Zhang J, Wang J, Xu J, Lu Y, Jiang J, Wang L, Shen H-M, Xia D (2016) Curcumin targets the TFEB-lysosome pathway for induction of autophagy. Oncotarget 7(46):75659–75671
75. Moustapha A, Pérétout PA, Rainey NE, Sureau F, Geze M, Petit J-M, Dewailly E, Slomianny C, Petit PX (2015) Curcumin induces crosstalk between autophagy and apoptosis mediated by calcium release from the endoplasmic reticulum, lysosomal destabilization and mitochondrial events. Cell Death Dis 1:15017
76. Leslie EM, Deeley RG, Cole SPC (2005) Multidrug resistance proteins: role of P-glycoprotein, MRP1, MRP2, and BCRP (ABCG2) in tissue defense. Toxicol Appl Pharmacol 204 (3):216–237
77. Mulik RS, Mönkkönen J, Juvonen RO, Mahadik KR, Paradkar AR (2012) Apoptosis-induced anticancer effect of transferrin-conjugated solid lipid nanoparticles of curcumin. Cancer Nanotechnol 3(1–6):65–81
78. Dhule SS, Penfornis P, Frazier T, Walker R, Feldman J, Tan G, He J, Alb A, John V, Pochampally R (2012) Curcumin-loaded γ-cyclodextrin liposomal nanoparticles as delivery vehicles for osteosarcoma. Nanomedicine 8(4):440–451
79. Zhao Y, Huan M, Liu M, Cheng Y, Sun Y, Cui H, Liu D, Mei Q, Zhou S (2016) Doxorubicin and resveratrol co-delivery nanoparticle to overcome doxorubicin resistance. Sci Rep 6 (1):1–15
80. Wang S, Ren W, Liu J, Lahat G, Torres K, Lopez G, Lazar AJ, Hayes-Jordan A, Liu K, Bankson J, Hazle JD, Lev D (2010) TRAIL and doxorubicin combination induces proapoptotic and antiangiogenic effects in soft tissue sarcoma in vivo. Clin Cancer Res 16 (9):2591–2604
81. Rahimi HR, Nedaeinia R, Sepehri Shamloo A, Nikdoust S, Kazemi OR (2016) Novel delivery system for natural products: nano-curcumin formulations. Avicenna J Phytomed 6(4):383–398

82. Almeida EAMS, Bellettini IC, Garcia FP, Farinácio MT, Nakamura CV, Rubira AF, Martins AF, Muniz EC (2017) Curcumin-loaded dual pH- and thermo-responsive magnetic microcarriers based on pectin maleate for drug delivery. Carbohydr Polym 171:259–266
83. Narayanan NK, Nargi D, Randolph C, Narayanan BA (2009) Liposome encapsulation of curcumin and resveratrol in combination reduces prostate cancer incidence in PTEN knockout mice. Int J Cancer 125(1):1–8
84. Nounou MI, ElAmrawy F, Ahmed N, Abdelraouf K, Goda S, Syed-Sha-Qhattal H (2015) Breast cancer: conventional diagnosis and treatment modalities and recent patents and technologies. Breast Cancer 9(Suppl 2):17–34
85. Wong HL, Bendayan R, Rauth AM, Xue HY, Babakhanian K, Wu XY (2006) A mechanistic study of enhanced doxorubicin uptake and retention in multidrug resistant breast cancer cells using a polymer-lipid hybrid nanoparticle system. J Pharmacol Exp Ther 317(3):1372–1381
86. Tanabe M, Ito Y, Tokudome N, Sugihara T, Miura H, Takahashi S, Seto Y, Iwase T, Hatake K (2009) Possible use of combination chemotherapy with mitomycin C and methotrexate for metastatic breast cancer pretreated with anthracycline and taxanes. Breast Cancer 16 (4):301–306
87. Sebak S, Mirzaei M, Malhotra M, Kulamarva A, Prakash S (2010) Human serum albumin nanoparticles as an efficient noscapine drug delivery system for potential use in breast cancer: preparation and in vitro analysis. Int J Nanomedicine 5:525–532
88. Chen Y, Zheng X-L, Fang D-L, Yang Y, Zhang J-K, Li H-L, Xu B, Lei Y, Ren K, Song X-R (2014) Dual agent loaded PLGA nanoparticles enhanced antitumor activity in a multidrug-resistant breast tumor xenograft model. Int J Mol Sci 15(2):2761–2772
89. Krey AK, Hahn FE (1969) Berberine: complex with DNA. Science 166(3906):755–757
90. Guo N, Yan A, Gao X, Chen Y, He X, Hu Z, Mi M, Tang X, Gou X (2014) Berberine sensitizes rapamycin-mediated human hepatoma cell death in vitro. Mol Med Rep 10 (6):3132–3138
91. Chen X, Zhang Y, Zhu Z, Liu H, Guo H, Xiong C, Xie K, Zhang X, Su S (2016) Protective effect of berberine on doxorubicin-induced acute hepatorenal toxicity in rats. Mol Med Rep 13 (5):3953–3960
92. Pan Y, Zhang F, Zhao Y, Shao D, Zheng X, Chen Y, He K, Li J, Chen L (2017) Berberine enhances chemosensitivity and induces apoptosis through dose-orchestrated AMPK signaling in breast cancer. J Cancer 8(9):1679–1689
93. Zhao Y, Jing Z, Li Y, Mao W (2016) Berberine in combination with cisplatin suppresses breast cancer cell growth through induction of DNA breaks and caspase-3-dependent apoptosis. Oncol Rep 36(1):567–572
94. Toivanen R, Shen MM (2017) Prostate organogenesis: tissue induction, hormonal regulation and cell type specification. Development 144(8):1382–1398
95. Jemal A, Center MM, DeSantis C, Ward EM (2010) Global patterns of cancer incidence and mortality rates and trends. Cancer Epidemiol Biomark Prev 19(8):1893–1907
96. Ammon HP, Wahl MA (1991) Pharmacology of Curcuma longa. Planta Med 57(1):1–7
97. Li L, Ahmed B, Mehta K, Kurzrock R (2007) Liposomal curcumin with and without oxaliplatin: effects on cell growth, apoptosis, and angiogenesis in colorectal cancer. Mol Cancer Ther 6(4):1276–1282
98. Li L, Braiteh FS, Kurzrock R (2005) Liposome-encapsulated curcumin: in vitro and in vivo effects on proliferation, apoptosis, signaling, and angiogenesis. Cancer 104(6):1322–1331
99. Sanna V, Siddiqui IA, Sechi M, Mukhtar H (2013) Resveratrol-loaded nanoparticles based on poly(epsilon-caprolactone) and poly(D,L-lactic-co-glycolic acid)-poly(ethylene glycol) blend for prostate cancer treatment. Mol Pharm 10(10):3871–3881
100. Harper CE, Patel BB, Wang J, Arabshahi A, Eltoum IA, Lamartiniere CA (2007) Resveratrol suppresses prostate cancer progression in transgenic mice. Carcinogenesis 28(9):1946–1953

Pharmacokinetics, Interaction, and Toxicological Profile of Nanophytomedicine

Vineet Kumar Rai, Raj Kumar Narang, Faheem Hyder Pottoo, and Md Abul Barkat

Abstract

Pharmacokinetic parameters are proficiently altered to get the desired pharmacokinetic profile of the drugs and reduce the toxic potential of the loaded drugs. As per clinical trials done till 2016, as much as 78 nanocarrier-based formulations were available in the pharmaceutical market globally of which 63 nanomedicines were approved. It is anticipated that >60% of the new drug molecules have very less aqueous solubility, hampering advancement, and accomplishment of newer treatment options. Various nanocarrier-based approaches like nanoparticles (polymeric or metallic) and nanoemulsions have been developed to encapsulate an active molecule, an antibody, a detection probe, and/or genes. The ADME of the drug loaded in nanoformulations can differ extensively based on the nature of drug delivery carrier opted, the properties of the nanomedicines and the unevenness amid organ size, body fat index, etc. of different individuals. However, nanomedicines have shown important uniqueness like the way of transport through the fine blood capillaries and lymphatic endothelium, ability to prolong circulation time and increase blood level, increase binding ability with the biomolecules (like proteins), more accumulation at the site of target tissues, and alleviated inflammatory or immune responses. The importance of particle size of the developed nanomedicines is also justified for lengthening of drug stay in specific target site, and to eliminate them quickly once delivered to the non-specific sites. Therefore, ultimate aim of developing nanomedicine is to

V. K. Rai (✉) · R. K. Narang
Department of Pharmaceutics, ISF College of Pharmacy, Moga, Punjab, India

F. H. Pottoo
Department of Pharmacology, College of Clinical Pharmacy, Imam Abdulrahman Bin Faisal University, Dammam, Saudi Arabia

M. A. Barkat
Department of Pharmaceutics, College of Pharmacy, University of Hafr Al-Batin, Hafr Al-Batin, Saudi Arabia

S. Beg et al. (eds.), *Nanophytomedicine*,
https://doi.org/10.1007/978-981-15-4909-0_8

enhance its therapeutic efficacy of the loaded phytomedicines and to minimize systemic toxicity by improving their pharmacokinetic profile. This chapter however includes various aspects of improved pharmacokinetic profile, improved interaction with the biological membrane, and improved safety profile of the phytomedicine after loading inside nanostructured drug delivery systems.

Keywords

Nanostructured drug delivery systems · Phytomedicine · Pharmacokinetics · Biomolecules · Detection probes

1 Introduction

When drug enters into the biological environment of GIT, it follows a number of phases that are described by using the acronym LADME. Libration is another term that is being used today along with ADME i.e. meant for drug absorption, their distribution in the biological medium, metabolism, and elimination from the body. L in LADME means liberation of the active substance from the delivery system. It is also equally important as absorption. Liberation and absorption of the drug are cited as one phase in various literature. It is done when the drug is often administered in an active form and following no release phase out there before absorption. However, absorption is completely meaningless unless libration or release of the drug from a delivery system is justified [1, 2]. As far as nanophytomedicine is concerned, liberation step should be justified firstly before making any pharmacokinetic profile of the dosage forms. One more acronym is used i.e. ADME-Tox or ADMET that reflect drug's toxicological aspect. Most of the nanophytomedicines are meant to control the release of the API [3, 4]. Site-specific drug delivery and controlling release of drug from systems prolong drug staying inside the body and hence greater tissue distribution and redistribution in case of cytotoxic drug can be observed [5]. Hence with reference to conventional dosage form which leads to relatively higher number of normal cells deaths of the body, nanomedicines offered selectivity towards infectious disease so it is another important aspect that is looked into [6–8]. Pharmacokinetic parameters are decided by the physicochemical characteristics of the drugs [1, 2]. These parameters are proficiently altered to get the desired pharmacokinetic profile of the drugs and reduced the toxic potential of the loaded drugs. As per clinical trials done till 2016, as much as 78 nanocarrier-based formulations were available in the pharmaceutical market globally of which 63 nanomedicines were approved (http://www.clinicaltrial.gov). It is anticipated that >60% of the new drug molecules have very less aqueous solubility, hampering advancement and accomplishment of newer treatment options [9, 10]. Subsequently, approaches for the management of various clinical disorders such as epilepsy and cancer are exhibiting poor therapeutic responses by monotherapy because of involving cross talk signaling and complex interfering mechanisms [12, 13]. Therefore, various approaches are being used to deliver synergistic combinations with the help of nanocarrier-based approaches [14, 15]. Nanomedicines comprise nanoparticles (polymeric or metallic) and nanoemulsions (liquid nanocarriers). Nanomedicines are developed to encapsulate active molecules, antibodies, detection probes, and various genes [5, 11].

Broadly explored advantages of nanomedicines as compared to that of the traditional medicines comprise their higher effectiveness, less toxicity, improved physicochemical, pharmacokinetic/ pharmacodynamic properties with the help of adding suitable pharmaceutical ingredients. In particular, a variety of kinetic profiles of nanomedicines in physiological conditions is modulated by making their formulations. It is but obvious that nanomedicines have a drastically dissimilar ADME profile in contrast to that of the conventional formulations. Some nanomedicines may improve the absorption of the drug molecules by releasing the drug in a controlled manner into gut and hence may solve the issues associated with that of the solubility. The intestinal wall is well designed (due to the presence of microvillus) for better absorption of the drugs and work as a barrier for the foreign invaders. Investigations have been done to develop nanomedicines with improved mucus-incisive abilities of the drugs [16]. The failure of some nanomedicines to endure proficient oral absorption demands another route for administration. Non-oral administrations may offer extra benefits like site-specific targeting to the intended site and show long term effects [17]. The skin proved as an attractive route for drug administration loaded in nanomedicine, as it may shun the issues related with i.v. administration away and the discomforts related to fluctuated gastric pH, altered gastric emptying time, and first-pass metabolism. Transdermal administration at the same time has widely been used for nanomedicines in the form of solid lipid nanoparticles (SLNs) and NEs. These systems have been distinguished as a biocompatible option with adequate safety and controlled drug release behavior [18, 19]. Topical ocular drug delivery offers a valuable administration route for nanomedicines treating ocular pathologies. Administration of nanomedicine based intraocular drugs to lower pressure [20] and antiapoptotic drugs [21] has successfully been accomplished in small animal models. Nasal administration of some nanomedicines has been investigated, keeping in mind that the nanomedicines efficiently penetrate nasal mucosa; for instance, chitosan nanoparticles of tizanidine increased brain availability of the drug due to increased penetration following nasal administration and finally increased drug efficacy in vivo [17]. So many investigations are carried out claiming the improvement of oral bioavailability of peptide and proteins. Once a drug encapsulating nanomedicine enters in the systemic circulation, the consequent delivery of the drug into biological tissues starts. The distribution of nanoformulations can differ extensively based on the nature of drug delivery carrier opted, the properties of the nanomedicines, and the unevenness amid organ size, body fat index, etc. of different individuals. Nanomedicines have shown important uniqueness like the way of transport through the fine blood capillaries and lymphatic endothelium, ability to prolong circulation time and increase blood level, increase binding ability with the biomolecules (like proteins), more accumulation at the site of target tissues, and alleviated inflammatory or immune responses. This distinctiveness varies from conventional medicines when checked on the basis of

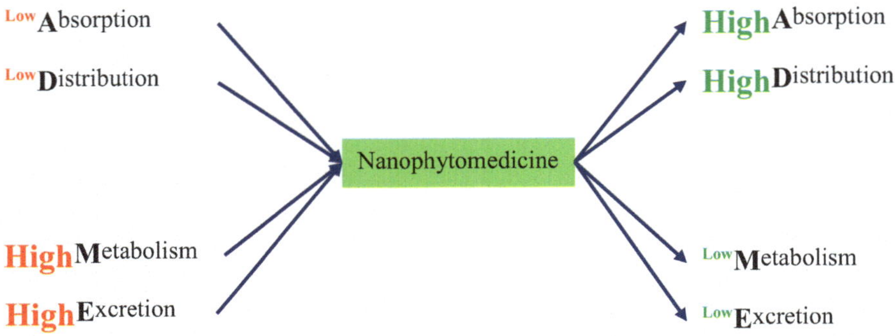

Fig. 1 Fate of nanophytomedicines in modulating pharmacokinetic parameters of the loaded phytoactives

physiochemical properties (particle surface, size, and chemical composition) of the nanomedicine [22–25]. Pharmacokinetic distinctiveness of several nanocarriers with diverse formulations are assessed in terms of particle size, shape, and surface characteristics. Nanocomposites with particle size <10 nm are eliminated from the kidneys; however, those with size >10 nm are circulated and eliminated from the liver and/or mononuclear-phagocyte system. The objective of modulation of particle size in nanomedicines is to lengthen their stay in specific target site, and to eliminate them quickly once delivered to the non-specific sites. Therefore, the ultimate aim of developing nanomedicine is to enhance its therapeutic efficacy of the loaded phytomedicines and to minimize systemic toxicity by improving their pharmacokinetic profile. This chapter however includes various aspects of improved pharmacokinetic profile, improved interaction with the biological membrane, and improved safety profile of the phytomedicine after loading inside nanostructured drug delivery systems (Fig. 1).

As far as drug-loaded nanocarriers are concerned, pharmacokinetic profile of the drug should include libration of the drug as a primary parameter in addition to absorption, distribution, metabolism, and excretion. The acronym may be represented for the pharmacokinetic study of the drug loaded in nanocarriers by LADME. However, this acronym cannot be completed unless toxicity related issues of the drug are discussed. Hence the final acronym should be termed as LADME-Tox. Various types of nanophytomedicines have been investigated for improved release/libration of the drug, absorption, widespread distribution, delayed metabolism and excretion and reduced toxicity of the drug due to unwanted site-specificity. Some of the examples are presented in Table 1.

Table 1 List of nanocarriers developed to improve the pharmacokinetic profile of the phytomedicines

Novel carrier system	Drug	Indication	Targeted pharmacokinetic modulations	Major outcome	References
Stealth, biocompatible monoolein-based lyotropic liquid crystalline PEGylated nanoparticles	Aloe-emodin	Cancer (in vitro)	Improved aqueous solubility, intestinal absorption, delayed pre-systemic metabolism via glucuronidation and P-glycoprotein efflux process	• ~96% drug encapsulation efficiency • Good serum stability • Hemocompatibility • Higher cytotoxic effects. • Higher cellular uptake in MCF-7 cells. • PEGylated AE-LCNPs demonstrated increased bioavailability of AE compared with free AE	[37]
Liposome	Baicalin	Cerebral ischemia (in vivo)	Improved drug lipophilicity for enhanced drug absorption in the brain tissues	• Sustained release of the drug • Prolong drug retention time in vivo • Increased the drug concentration in the brain • Increased drug concentration in the brain	[26]
Lecithin-stabilized Micellar drug delivery system (LsbMDDs)	Docetaxel	Cancer	Enhanced absorption and distribution Prolonged elimination	• High drug loading • Enhanced efficacy against cancer • Reduced systemic toxicity	[38]
Solid lipid nanoparticles for Oral delivery	Houttuynia cordata extract for diuresis and detoxification	Antiviral and antibacterial activities	Taste masking Improved water solubility and oral bioavailability	• High drug loading • Controlled release • Specific Caco-2 cells death	[39]
Polyethyleneglycol-co-polycaprolactone-co-polyethylenimine nanoparticles	Rhein	Diabetic nephropathy	Improved solubility, bioavailability, kidney distribution, and reduced adverse effects	• Improved kidney distribution • Reversing DN • Improved pharmacological effects targeted drug delivery	[40]

(continued)

Table 1 (continued)

Novel carrier system	Drug	Indication	Targeted pharmacokinetic modulations	Major outcome	References
Starch nano-spheres	Ginkgo biloba extracts	Memory impairment and Alzheimer's, dementia	Improved bioavailability	• Controlled release of loaded extracts from nanocarriers • Widen application in the field of biopharmaceutics and the functional food industry	[41]
PCL/PTMC composite fiber mats	Shikonin	Healing of wounds, scar, and Burns	Widen biomedical applications	• Along with the free radical scavenging activity, the antibacterial effects of the shikonin-loaded fiber mats indicated that it could act as wound healer and antibacterial agent	[42]
Self-assembled licorice protein nanoparticles	Aconitine	Toxicity	Reduced toxicity	• Showed detoxifying effects of licorice against aconite • Provides a new approach and material for application of active phytochemicals and herbal compositions with much less safety concerns	[43]
Polymeric nanoparticles	Curcumin	Toxicity	Protection from carbon tetrachloride (CCl_4)-induced subacute hepatotoxicity	• Hepatoprotective effect at relatively low dose of CurNP due to the encapsulation of cur in nanoparticles, sustained release, better circulation kinetics and ability to target liver	[44]

Nanoparticles	Bismuth	Cancer	Cytotoxicity against HT-29 cell line	• Bismuth nanoparticle induced cytotoxicity on HT-29 cell line through the activation of late apoptosis independent of caspase pathway	[45]

2 Pharmacokinetics of Nanophytomedicine

Investigations are gradually providing evidence to present new phytoactives with therapeutic benefits based on their traditional use as plant based medicines. Clinical utility of a herbal drug with the promising therapeutic effect is handicapped due to low aqueous solubility. Another point of concern is that their abruption from the biological membrane. Lesser solubility led to the lower level of drug concentration for absorption to the systemic circulation. Direct availability of the drugs at the site of absorption ensures a higher pool of drug for better absorption. Improvement in absorption potential is directly related to the improved dissolution of the drugs and hence Cmax, Tmax, AUC all get improved significantly [25]. Availability of the active drug in solubilized form, thickness of the mucous membrane, logP, and many more factors affect the absorption of the drug through the biological membrane. Distribution of drug in systemic circulation and organs depends on plasma protein binding of the drug, biological half-life, stability of the compound at specific pH or temperature, etc. Metabolic enzymes like cytochrome P450, glucuronosyltransferase enzymes may facilitate or hamper drug metabolism. Rate of excretion depends on the availability of metabolized drug and the volume of distribution in the systemic circulation. Pharmacokinetic properties get affected upon following different route of administration and amount of drug administered. Nanophytomedicines are prepared to alter the pharmacokinetic pattern of the actives of choice. Encapsulation directly involves surface modification of the drug and hence modifies the rate of absorption of the less absorbed actives. List of the nanocarriers developed to modulate the pharmacokinetic profile of the phytomedicine is given in Table 1.

History of using herbs as medicine is thousands of years old. Roots of *Scutellaria baicalensis* has been used as traditional medicine in China. It has been investigated to be effective against cerebral ischemia due to the presence of Baicalin as the main component. In addition, it has been investigated active against various other diseases such as infectious diseases, brain edema, and cancer. Furthermore, radical scavenging activity and neuroprotective effects from ischemia–reperfusion injury in vivo are also reported; however, due to its less lipid and aqueous solubility, it has poor permeation through the blood–brain barrier. It is the main limitation in the delivery of Baicalin to the brain for efficient therapeutic effect. In order to improve its clinical application in the management of the neuro-disorders, it was loaded inside a novel drug delivery system. By this approach, its solubility was improved significantly. Baicalin-loaded liposomes enhanced the lipophilic nature of the drugs and the drug level in brain tissues by modulating their pharmacokinetic properties. Liposomes are efficient nanosize lipid carriers that can encapsulate all hydrophilic, lipophilic, and amphoteric molecules due to the presence of both lipophilic bilayer and hydrophilic core in it. Phospholipids to drug ratio 3.81:1, phospholipid to cholesterol proportion 5.70:1, aqueous part 1.02 mL, gave spherical particles with an average size of approximately 175 nm and negative surface charge. The release of BA from these liposomes was slow that demonstrated constant and extended drug release from the carriers. Pharmacokinetic investigations in animals resulted the effect of stroke on disposition and absorption of BA through liposomes. MCAO group gives higher values of Cmax and AUC0–t as compared to the normal animals. In the pharmacokinetics investigation on MCAO rat's plasma, the concentration maxima and area under the curve values were observed to be considerably higher with BA-LP than that of BA. About 2.87-times improvement in the $t_{1/2}$ and 8.08-times alleviation in the renal clearance rate with a 2.14-times improvement in the MRT0–t on MCAO model show the proficiency of BA-LP over BA. The tissue distribution of BA was observed to be considerably changed in the case of BA-LP as compared to that of BA alone. The level of presence of BA in the heart, liver, spleen, lungs, and brain was observed to be augmented following BA-LP administration in contrast with that of BA in both normal and MCAO models, however, alleviated in the kidneys [26].

Several investigations have been done showing improved oral bioavailability and therapeutic effectiveness of pharmaceuticals and nutraceuticals when delivered in the form of nanophytomedicines. The nanosized formulation approach may help improve maximum drug plasma level and bioavailability up to 12 times in contrast to the traditional dosage forms with micrometer particle size. Fascinatingly, the neutral or acidic molecules like danazol, cilostazol, tranilast, and curcumin have been investigated to show significant improvements in the pharmacokinetic parameters in nanocrystal technology. Lipid nanocarrier has been successfully used to provide sustained release of docetaxel for up to 72 hours. Nanoliposomes have shown 2.14 folds improvement in basolateral docetaxel internalization in the Caco-2 cell monolayer as compared to that of its solution counterpart and about 3.65 folds improvement in the oral bioavailability of the docetaxel than its solution. DTX-LNs have been proved efficient to improve the oral absorption of docetaxel [27]. Intrinsic behaviors of the nanomaterials have shown several benefits in pharmaceutical product development. The small size of nanomaterials has a very explicit surface area in relation to the volume, because of which the particle surface energy is improved that makes nanomaterials more imprudent. Nanomedicines help accumulation of actives at the target site and lower therapeutic doses. Nanoparticles efficiently used to modulate physicochemical characteristics of actives that retard them to cross blood–brain barrier. After crossing blood–brain barriers, these systems are meant to provide sustained release, increase bioavailability, and avoid peripheral toxicity. Nanoparticles have been used as the most promising technology for the delivery of therapeutic molecules transversely to the BBB. Specially designed nanoparticles of optimal size and surface characteristics have been studied to enhance the circulation time of the herbal drugs in the bloodstream and to increase the biodistribution of the chemotherapeutic agents. Nanotechnologies of the next generation are meant to have a multifunctional approach and are planned in a multiplex manner. These are being specially designed to facilitate personalized and tailored cancer treatment.

Metabolism in the biological system is one of the important pharmacokinetic parameters of the drug. Pre-uptake metabolism within the GI tract leads to poor oral bioavailability and that too is eliminated using a nanotechnology approach. Escaping of hepatic metabolism, degradation at gastric pH, enzymatic degradation are some of the metabolism related issues of the plant based drugs that have been solved using nanotechnology based approaches. Among investigations carried out for the advancement of the novel nanocarriers, lipid nanoparticles (SLN and NLCs) have shown immense utility in site-specific drug delivery. For the treatment of disorders associated with central nervous systems, the nasal route of administration has shown proficiency. Various types of drug-loaded lipid nanocarriers have particularly been formulated for the treatment of neurodegenerative disorders. This approach has been considered suitably efficient due to the direct brain drug delivery by the nasal route. The feasibility of this approach opened a new vista of pharmaceutical application for investigators working in the arena of drug delivery. This approach has been used to overcome several drawbacks (low absorption and rapid mucociliary clearance) that resist clinical pertinence of pharmaceutical actives. This approach has been used to improve the systemic absorption of therapeutically active moieties against cardiovascular diseases, infections, and severe pain. These nanotechnologies are specifically meant to deliver phytoactives through bypassing BBB, bypassing hepatic metabolism and gastrointestinal degradation [28]. Elimination is the final process that is investigated in the ADME process of pharmacokinetics. Elimination of the drug or ingredients depends on various factors. Degradation pattern of nanocarriers in response to stimuli of a suitable environment, mucus permeation, enzyme inhibitory properties, and their toxicity due to non-clearance are crucial factors that are taken into consideration while designing nanocarriers. Renal or biliary elimination are also crucially analyzed before designing these nanocarriers. Degradation kinetics is a significant parameter that extends drug release and makes difficult the designing of drug delivery carriers with predetermined drug release behavior. The body's defense system is based on the immunity of the body. It is meant to remove foreign objects from the body whether it is microbe or nanocarriers. Immunogenicity is the factor that is sometimes associated with herbal actives. This situation needs avoidance of direct exposure to active with the immune systems. The process of avoidance of immunogenicity using any external method is called opsonization. PEGylated nanoparticles have immensely been used for opsonization. This process largely regulates the process of accelerated blood clearance as these help modify the surface of the drug molecule. As per the Kadam Rajendra et al. report, entrapment of phytomolecules in nanocarriers alleviates the noticeable drug clearance from plasma, thereby increasing the noticeable drug passage biological half-life and potential cumulative drug delivery to the specific site. In addition, positive surfaced nanocarriers have extended noticeable biological half-life and alleviated plasma clearance than the neutral and/or negatively charged nanoparticles, resulting in efficient drug delivery. Besides alleviated clearance of the gastrointestinal tract's lumen, liver, and from circulation, it is expected that the alleviated clearance from the administration or absorption site may also give improved drug exposure with nanoparticles [29].

HPMC and PVP combination has been used to prepare hydrogel nanoparticle of curcumin to improve its absorption and prolong the rapid clearance owing to probable avoidance by the reticuloendothelial system. These nanoparticles with 100 nm size, homogeneous distribution, and 72% entrapment effectiveness resulted in improved curcumin absorption, improved in vivo anti-malarial effect. Acute and subacute toxicity investigations showed the oral safety of the formulation [30]. At the same time, nanocrystals grasp immense ability as parenteral delivery carrier systems for less aqueous soluble APIs. Explanation of the *in vivo* relevance of parenteral nanocrystals is of pharmacological, toxicological, and mechanistic importance. However, it is very tough to examine concurrent imaging of nanocrystals *in vivo* because they are successively dissolve in the biological system. Self-discriminating hybrid nanocrystals (SDHNs) of curcumin have been prepared by matrix environment-sensitive fluorescent dyes traces into the crystalline network of curcumin. In IV administration, the major portion of CUR nanocrystals are eliminated from plasma speedily and some of them are collected by liver and lung. Prolonged circulation of the drug by nanocrystals is attributed to the poloxamer 188 coating and use of suitable stabilizers during preparation. However, the final relevance of the nanocrystals was to avoid reticuloendothelial cells present in different organs. They observed that the parenteral delivery provided controlled release also due to which, pharmacological efficacy was observed for an extended period. However, it raises fear of toxicity in vital organs and tissues in parallel, particularly when the drugs are very toxic [31].

3 Toxicity of Nanophytomedicine

Application of nanotechnologies has revolutionized various aspects of technology. The evolution of nanotechnology led to widening the application of phytoactives by facilitating the development of efficient drug delivery systems, sunscreens dosage forms, cosmetic preparations, antibacterial formulations, and many more. Nanocarriers offer exclusive physicochemical characters due to the high surface area, surface and volume proportion, small size range, and diverse optical behaviors [32]. Nanotechnologies, in addition to their medical application, have become an important tool for diverse applications in the field of oncology, diagnosis, and drug delivery. Nanoparticles are expected to reduce associated toxicity of chemotherapeutic agents and other actives having a narrow therapeutic window as many in vitro and *in vivo* studies demonstrate their toxic effects in biological systems like cellular, organ, and tissue and also offer some allergic responses and inflammations [33]. The investigations carried out for nanomedicine/nanophytomedicine are based mainly on the clinical pertinence; however, clinical pertinence of these carriers is highly dependent on the toxicity related issues of the nanocarriers. These issues are not usually taken into consideration [34]. The information about the toxic potential of nanophytomedicine is less explored; however, it is increasing at a fast pace. List of toxicity cases associated with nanophytomedicines is presented in Table 2.

Titanium dioxide nanoparticles (TiO2NPs) are the most frequently used NPs in paints, papers, plastics, sunscreens, medicines, foods, cosmetics, and toothpaste. These are extensively used owing to their high degree of stability, anticorrosive, and photocatalytic properties. However, knowing the toxicological effects of TiO2NPs is equally important to widen their professional and industrial use. Investigating their toxicology is an important issue as these nanoparticles may easily get entered in the human body through inhalation, ingestion, and dermal penetration. These NPs may

Table 2 List of toxicity cases associated with nanophytomedicines

Novel carrier system	Properties of the carrier	Toxicity case	References
Titanium dioxide nanoparticles (TiO2NPs)	High degree of stability, anticorrosive, and photocatalytic	After inhalation, ingestion, and dermal penetration, these NPs may cause oxidative stress, genotoxicity, cytotoxicity, inflammation, and cell apoptosis in mammal organs and systems.	[46]
Cationic nanobubbles	Due to the presence of hexadecyltrimethyl–ammonium bromide (CTAB) positive charge is developed	Influence cell proliferation, inflammation and differentiation, and production of pro-apoptotic genes in epithelial cells	[35]
Lipid and polymeric nanoparticles	170–225 nm particle size and positive surface charge	Unlike neutral charged nanoparticles, cationic nanoparticles started apoptosis and release of the mediators with higher toxic effect. This is due to the increased interaction with the cell membrane. It also resulted in programmed cell death of the normal cell.	[33]
Needle-shaped carbon nanotubes, nanowires, and nanofibers	After being inhaled, micron-sized particles get accumulated in the central and peripheral lung airways	Fibrotic lung disease and rare tumors	[24]
Metal nanoparticles i.e. SiONPs primed with lipopolysaccharide	Acute pulmonary effect of metal nanoparticles	Pro-inflammatory cytokines and inflammatory cell infiltration in the broncho-alveolar lavage fluid. Lipid peroxidation, oxidative stress, oxidative DNA damage, disruption of the cell membrane, programmed cell death, and reduced proliferation	[36]

cause oxidative stress, genotoxicity, cytotoxicity, inflammation, and cell apoptosis basically in mammal organs and systems. These NPs are observed to cause body, organs, tissues, and cellular level toxicity. These have potential health risks of cardiovascular diseases [32]. Nanocarriers are inclined to generate reactive oxygen species and free radicals, offering oxidative stress and inflammation like conditions along with DNA damage and fibrosis. Accumulation inside the cells is another point of apprehension in nanodrugs application. Nanocarriers between the size range of 60 and 200 nm are most likely to produce toxicity. Nanocarrier associated toxicities are also depended on their other physical and chemical prospects like shape and surface charge, area, and extent of reactivity. Cationic surface charge over the nanocarrier surface has systematically been portrayed to influence cell proliferation and differentiation, and production of pro-apoptotic genes in epithelial cells. The polycationic character of the nanoformulations provokes necrosis and apoptosis. Cationic surfactants are sometimes used in the development of nanoparticles; therefore, use of cationic surfactants in nanoparticular drug delivery systems has fully been elucidated by evaluating inflammatory effects of cationic nanobubbles. Due to the presence of hexadecyltrimethyl–ammonium bromide (CTAB), zeta potential of nanobubbles was reached to +80 mV that resulted in cytotoxic effects on neutrophils in a concentration-dependent manner [35].

Additionally, the comparative toxic effect of lipid and polymeric nanoparticles on human neutrophils in the existence of cationic surfactants has also been investigated on the basis of cytotoxic potential and level of induction of pro-inflammatory mediators like lactate dehydrogenase (LDH), elastase, $O2^{\bullet -}$, and intracellular Ca^{2+}. The size of these nanoparticles was kept within the range of 170–225 nm and positive charge over the particles surface was ensured using cetyltrimethylammonium bromide or soyaethyl morpholinium ethosulfate i.e. the cationic surfactant. Nanoparticles with no positive charge resulted in significant immune and inflammatory responses; however, cationic surfactants in both types of nanoparticles and free forms started apoptosis and efficient release of the mediators with higher toxic effect in the case of lipid nanoparticles. Cationic nanoparticles interact with the cell membrane due to the cationic charge triggering membrane disruption followed by rapid Ca^{2+} influx. Higher intracellular Ca^{2+} level induces degranulation and leads to oxidative stress as a result programmed cell death of the normal cell starts. Therefore, the selection of cationic additives in the development process of nanocarriers should cautiously be completed [33]. The surface effect of the produced nanoparticles may produce chemical harm to the nearby tissues. After being inhaled, micron-sized particles be inclined to get accumulated in the central airways; however, inhaled nanoparticles be able to accumulate in the peripheral lung producing inflammation of high degree. Particularly, needle-shaped carbon nanotubes, nanowires, and nanofibers could produce fibrotic lung disease and rare tumors [24]. The dermal penetration of nanoparticles is based on the physical and chemical perspective of the nanoparticles and skin situation. Contrary to this benevolent conclusion, many nanoparticles have been investigated to be toxic to the cells and pro-inflammatory to in vitro skin cell lines. As a result, supplementary investigation of chronic dosing of nanodrugs will be required in small animal models for subsequent support of nanotoxicity. Liver and spleen are the major organs that involve RES systems. The nanoparticles are commonly taken through RES systems. Nanoparticles of biomedical use are coated with biocompatible materials to shun opsonization and RES uptake [33]. Most of the dendrimers show toxic and hemolytic effects owing to the presence of positive charge on their surface. However, anionic dendrimers with customized surfaces exhibit diminished hemolytic activity. Therefore, the nanoparticles surface also impacts their own distribution in organs. The nanoparticles clearance from the body is also based on size and surface of the nanocarriers. Very small nanoparticles (below 30 nm) cleared rapidly from renal excretion; however, nanoparticles with size greater than 200 nm are efficiently taken up by RES system present in spleen and bone marrow. Very novel field in toxicology has evolved that is termed as nanotoxicology. It deals with investigations required to deal with that how nanomaterials go through the cells and tissues, and the particular biodistribution, degradation, and excretion.

Metal based nanoparticles are most often used is an array of pharmaceutical and biomedical applications. Ko *et al.* have examined acute pulmonary effects following IN administration of SiONPs in small animals primed with or without lipopolysaccharide. As per their observations, SiONPs increases pro-inflammatory cytokines and inflammatory cell infiltration in the broncho-alveolar lavage fluid. These metal nanoparticles induce airway inflammation with an increase in the phosphorylation of mitogen-activated protein kinases. The proportion of the inflammatory cell reaction is initiated by elevating the amount of SiONPs in the acute pulmonary disease model produced by LPS. SiONPs exposure exacerbates inflammatory pulmonary diseases. Hence SiONPs exposure can be the potential cause of toxicity in humans with respiratory problems. SiONPs can cause various types of toxic potentials, together with lipid peroxidation, oxidative stress, oxidative DNA damage, disruption of the cell membrane, programmed cell death, and reduced proliferation. However, the major route of introduction for SiONPs is the respiratory tract; a pulmonary toxicity examination should definitely be carried out before claiming beneficial effect of these nanoparticles against pulmonary disorders.

4 Conclusion

The term pharmacokinetics is the investigation of ADME profile of the drug; however, the era of formulation development where various novel drug delivery systems are being developed, the term liberation of the drug from the developed nanopharmaceuticals has also become the important part of it. At the same time, term pharmacodynamic is also investigated. It is the science of response of the phytoactive on the body. It gets started once the phytoactives are therapeutically available into the systemic circulation. Moreover, it is well known fact that diversity comes with some nuisance. Major limitation associated with the current scenario of phytoactive delivery through nanoformulation is their toxicity in the biological system. Therefore, toxicity has also become a significant term as far as pharmacokinetics is concerned. The final investigational hypothesis cannot be fulfilled unless toxicity is investigated. Therefore, term LADMERT which is the combination of liberation, absorption, distribution, metabolism, excrition, response and toxicity, is the collective dose and response steps involved in any phytopharmacuticals delivery. It is the final term that formulation scientist should think of while exploring nanophytomedicine in near future.

References

1. Javed MN, Alam MS, Waziri A, Pottoo FH, Yadav AK, Hasnain MS, Almalki FA (2019) QbD applications for the development of nanopharmaceutical products. In: Pharmaceutical quality by design. Academic Press, Cambridge, pp 229–253
2. Javed MN, Kohli K, Amin S (2018) Risk assessment integrated QbD approach for development of optimized bicontinuous mucoadhesive limicubes for oral delivery of rosuvastatin. AAPS PharmSciTech 19(3):1377–1391
3. Sharma S, Javed MN, Pottoo FH, Rabbani SA, Barkat M, Sarafroz M, Amir M (2019) Bioresponse inspired nanomaterials for targeted drug and gene delivery. Pharm Nanotechnol 7(3):220–233
4. Mishra S, Sharma S, Javed MN, Pottoo FH, Barkat MA, Alam MS, Amir M, Sarafroz M (2019) Bioinspired nanocomposites: applications in disease diagnosis and treatment. Pharm Nanotechnol 7(3):206–219
5. Pottoo FH, Barkat MA, Ansari MA, Javed MN, Jamal QM, Kamal MA Nanotechnological based miRNA intervention in the therapeutic management of neuroblastoma. Semin Cancer Biol. https://doi.org/10.1016/j.semcancer.2019a.09.017
6. Alam MS, Javed MN, Pottoo FH, Waziri A, Almalki FA, Hasnain MS, Garg A, Saifullah MK (2019) QbD approached comparison of reaction mechanism in microwave synthesized gold nanoparticles and their superior catalytic role against hazardous nitro-dye. Appl Organomet Chem 33(9):e5071
7. Alam MS, Garg A, Pottoo FH, Saifullah MK, Tareq AI, Manzoor O, Mohsin M, Javed MN (2017) Gum ghatti mediated, one pot green synthesis of optimized gold nanoparticles: investigation of process-variables impact using box-Behnken based statistical design. Int J Biol Macromol 104:758–767
8. Hasnain MS, Javed MN, Alam MS, Rishishwar P, Rishishwar S, Ali S, Nayak AK, Beg S (2019) Purple heart plant leaves extract-mediated silver nanoparticle synthesis: optimization by box-Behnken design. Mater Sci Eng C 99:1105–1114
9. Sareen S, Mathew G, Joseph L (2012) Improvement in solubility of poor water-soluble drugs by solid dispersion. Int J Pharm Investig 2(1):12–17. https://doi.org/10.4103/2230-973X.96921
10. Sikarra DSV, Kharia AA, Chatterjee DP (2012) Techniques for solubility enhancement of poorly soluble drugs: an overview. J Med Pharm Allied Sci 01:1–22
11. Swarbrick J, Boylan JC (2002) Encyclopedia of pharmaceutical technology, vol 2. Marcel Dekker, New York
12. Pottoo FH, Tabassum N, Javed MN, Nigar S, Sharma S, Barkat MA, Alam MS, Ansari MA, Barreto GE, Ashraf GM (2020) Raloxifene potentiates the effect of fluoxetine against maximal electroshock induced seizures in mice. Eur J Pharm Sci 146:105261. https://doi.org/10.1016/j.ejps.2020.105261
13. Pottoo FH, Javed M, Barkat M, Alam M, Nowshehri JA, Alshayban DM, Ansari MA (2019) Estrogen and serotonin: complexity of interactions and implications for epileptic seizures and epileptogenesis. Curr Neuropharmacol 17(3):214–231
14. Pottoo FH, Tabassum N, Javed MN, Nigar S, Rasheed R, Khan A, Barkat MA, Alam MS, Maqbool A, Ansari MA, Barreto GE (2019) The synergistic effect of raloxifene, fluoxetine, and bromocriptine protects against pilocarpine-induced status epilepticus and temporal lobe epilepsy. Mol Neurobiol 56(2):1233–1247
15. Nigar S, Pottoo FH, Tabassum N, Verma SK, Javed MN (2016) Molecular insights into the role of inflammation and oxidative stress in epilepsy. J Adv Med Pharm Sci 10(1):1–9
16. Ensign LM, Schneider C, Suk JS, Cone R, Hanes J (2012) Mucus penetrating nanoparticles: biophysical tool and method of drug and gene delivery. Adv Mater 24(28):3887–3894
17. Patel D, Naik S, Misra A (2012) Improved Transnasal transport and brain uptake of Tizanidine HCl-loaded Thiolated chitosan nanoparticles for alleviation of pain. J Pharm Sci 101 (2):690–706. https://doi.org/10.1002/jps.22780

18. Garg GP, Prakash M (2002) In: Swarbrick J, Boylan JC (eds) Encyclopaedia of pharmaceutical technology, 2nd edn. Marcel Dekker, New York. https://doi.org/10.1021/op0340356. 2002. 3 Volumes. 3032 + 64 pp. ISBN 0-8247-2822-X (Vol. 1), 0-8247-2823-8 (Vol. 2), 0-8247-2824-6 (Vol. 3), 0-8247-2825-4 (Prepack), 0-8247-2825-4 (Online). Organic Process Research & Development. 2003;7(4):609-10
19. Khurana S, Jain NK, Bedi PMS (2013) Nanoemulsion based gel for transdermal delivery of meloxicam: Physico-chemical, mechanistic investigation. Life Sci 92(6):383–392. https://doi.org/10.1016/j.lfs.2013.01.005
20. Hathout RM, Mansour S, Mortada ND, Guinedi AS (2007) Liposomes as an ocular delivery system for acetazolamide: in vitro and in vivo studies. AAPS PharmSciTech 8(1):E1–E12. https://doi.org/10.1208/pt0801001
21. Nkansah MK, Tzeng SY, Holdt AM, Lavik EB (2008) Poly(lactic-co-glycolic acid) nanospheres and microspheres for short- and long-term delivery of bioactive ciliary neurotrophic factor. Biotechnol Bioeng 100(5):1010–1019. https://doi.org/10.1002/bit.21822
22. De Jong WH, Borm PJA (2008) Drug delivery and nanoparticles: applications and hazards. Int J Nanomedicine 3(2):133–149
23. Liu W, Yang X-L, Winston Ho WS (2011) Preparation of uniform-sized multiple emulsions and micro/Nano particulates for drug delivery by membrane emulsification. J Pharm Sci 100 (1):75–93. https://doi.org/10.1002/jps.22272
24. Onoue S, Yamada S, Chan H-K (2014) Nanodrugs: pharmacokinetics and safety. Int J Nanomedicine 9:1025–1037. https://doi.org/10.2147/IJN.S38378
25. Benchimol MJ, Bourne D, Moghimi SM, Simberg D (2019) Pharmacokinetic analysis reveals limitations and opportunities for nanomedicine targeting of endothelial and extravascular compartments of tumours. J Drug Target 27(5–6):690–698. https://doi.org/10.1080/1061186X.2019.1566339
26. Nan Li LF, Tan Y, Xiang Y, Zhang R, Yang M (2018) Preparation, characterization, pharmacokinetics and biodistribution of baicalin-loaded liposome on cerebral ischemia-reperfusion after i.v. administration in rats. Molecules 23:17–47
27. Hu K, Cao S, Hu F, Feng J (2012) Enhanced oral bioavailability of docetaxel by lecithin nanoparticles: preparation, in vitro, and in vivo evaluation. Int J Nanomedicine 7:3537–3545. https://doi.org/10.2147/IJN.S32880
28. Cunha S, Amaral MH, Lobo JMS, Silva AC (2017) Lipid nanoparticles for nasal/intranasal drug delivery. Crit Rev Ther Drug Carrier Syst 34(3):257–282. https://doi.org/10.1615/CritRevTherDrugCarrierSyst.2017018693
29. Kadam RS, Bourne DW, Kompella UB (2012) Nano-advantage in enhanced drug delivery with biodegradable nanoparticles: contribution of reduced clearance. Drug Metab Dispos 40 (7):1380–1388. https://doi.org/10.1124/dmd.112.044925
30. Dandekar PP, Jain R, Patil S, Dhumal R, Tiwari D, Sharma S et al (2010) Curcumin-loaded hydrogel nanoparticles: application in anti-malarial therapy and toxicological evaluation. J Pharm Sci 99(12):4992–5010. https://doi.org/10.1002/jps.22191
31. Wang T, Qi J, Ding N, Dong X, Zhao W, Lu Y et al (2018) Tracking translocation of self-discriminating curcumin hybrid nanocrystals following intravenous delivery. Int J Pharm 546 (1–2):10–19. https://doi.org/10.1016/j.ijpharm.2018.05.020
32. El-Din EAA, Mostafa HE, Samak MA, Mohamed EM, El-Shafei DA (2019) Could curcumin ameliorate titanium dioxide nanoparticles effect on the heart? A histopathological, immunohistochemical, and genotoxic study. Environ Sci Pollut Res Int 26(21):21556–21564. https://doi.org/10.1007/s11356-019-05433-2
33. Wick P, Manser P, Limbach LK, Dettlaff-Weglikowska U, Krumeich F, Roth S et al (2007) The degree and kind of agglomeration affect carbon nanotube cytotoxicity. Toxicol Lett 168 (2):121–131. https://doi.org/10.1016/j.toxlet.2006.08.019
34. Hwang TL, Sung CT, Aljuffali IA, Chang YT, Fang JY (2014) Cationic surfactants in the form of nanoparticles and micelles elicit different human neutrophil responses: a toxicological study. Colloids Surf B Biointerfaces 114:334–341. https://doi.org/10.1016/j.colsurfb.2013.10.021

35. Hwang T-L, Sung CT, Aljuffali IA, Chang Y-T, Fang J-Y (2014) Cationic surfactants in the form of nanoparticles and micelles elicit different human neutrophil responses: a toxicological study. Colloids Surf B Biointerfaces 114:334–341. https://doi.org/10.1016/j.colsurfb.2013.10.021
36. Ko J-W, Lee H-J, Shin N-R, Seo Y-S, Kim S-H, Shin I-S et al (2018) Silicon dioxide nanoparticles enhance endotoxin-induced lung injury in mice. Molecules (Basel, Switzerland) 23(9):2247. https://doi.org/10.3390/molecules23092247
37. May S, Freag YSE, Abdelmonsif DA, Abdallah OY (2016) Stealth, biocompatible monoolein-based lyotropic liquid crystalline nanoparticles for enhanced aloe-emodin delivery to breast cancer cells: in vitro and in vivo studies. Int J Nanomedicine 11:4799–4818
38. Su C-Y, Liu J-J, Ho Y-S, Huang Y-Y, Chang VH-S, Liu D-Z et al (2018) Development and characterization of docetaxel-loaded lecithin-stabilized micellar drug delivery system (LsbMDDs) for improving the therapeutic efficacy and reducing systemic toxicity. Eur J Pharm Biopharm 123:9–19. https://doi.org/10.1016/j.ejpb.2017.11.006
39. Ju-Heon Kim J-SB, Park J-K, Lee B-J, Kim M-S, Hwang S-J, Lee J-Y, Cho C-W (2017) Development of houttuyniacordata extract-loaded solid lipid nanoparticles for oral delivery: high drug loading efficiency and controlled release. Molecules 22:2215
40. Danfei Chen SH, Zhu Y, Hu F, Wei Y, Wang G (2018) Kidney-targeted drug delivery via rhein-loaded polyethyleneglycol-co-polycaprolactone-co-polyethyleneimine nanoparticles for diabetic nephropathy therapy. Int J Nanomedicine 13:3507–3527
41. Wang T, Wu C, Fan G, Li T, Gong H, Cao F (2018) Ginkgo biloba extracts-loaded starch nanospheres: preparation, characterization, and in vitro release kinetics. Int J Biol Macromol 106:148–157. https://doi.org/10.1016/j.ijbiomac.2017.08.012
42. Han J, Chen T-X, Branford-White CJ, Zhu L-M (2009) Electrospun shikonin-loaded PCL/PTMC composite fiber mats with potential biomedical applications. Int J Pharm 382(1):215–221. https://doi.org/10.1016/j.ijpharm.2009.07.027
43. Ke L-J, Gao G-Z, Shen Y, Zhou J-W, Rao P-F (2015) Encapsulation of aconitine in self-assembled licorice protein nanoparticles reduces the toxicity in vivo. Nanoscale Res Lett 10(1):449. https://doi.org/10.1186/s11671-015-1155-1
44. Gregory Marslin JP, Qi S, Franklin G (2018) Oral delivery of curcumin polymeric nanoparticles ameliorates CCl4-induced subacute hepatotoxicity in Wistar rats. Polymers 10:541
45. MojtabaShakibaie HF, Ameri A, Adeli-Sardou M, Jafari M, Rahimi HR (2018) Cytotoxicity of biologically synthesised bismuth nanoparticles against HT-29 cell line. IET Nanobiotechnol 12(5):653–657
46. El-Din EAA, Mostafa HE-S, Samak MA, Mohamed EM, El-Shafei DA (2019) Could curcumin ameliorate titanium dioxide nanoparticles effect on the heart? A histopathological, immunohistochemical, and genotoxic study. Environ Sci Pollut Res Int 26(21):21556–21564. https://doi.org/10.1007/s11356-019-05433-2

Recent Advancement in Clinical Application of Nanotechnological Approached Targeted Delivery of Herbal Drugs

Md Noushad Javed, Ekta Singh Dahiya, Abdallah Mohammad Ibrahim, Md. Sabir Alam, Firdos Alam Khan, and Faheem Hyder Pottoo

Abstract

Herbal medicinal practices are nowadays gaining popularity towards treatment of multiple acute and chronic diseases worldwide. In the past traditional formulations of herbal drugs in the form of juices, pastes, and decoction were exploited, but due to low shelf life of these preparations, herbs begin to be formulated in form of fermented syrups, tablets, capsules and medicated oils. But the active constituents of these herbal drugs such as, flavonoids, glycosides, terpenoids, are poorly absorbed from the gastrointestinal tract because of their

M. N. Javed
Department of Pharmaceutics, School of Pharmaceutical Education and Research - SPER (Formerly, Faculty of Pharmacy), Jamia Hamdard, New Delhi, India
e-mail: noushad.sch@jamiahamdard.ac.in

E. S. Dahiya (✉)
National Institute for Stroke and Applied Neurosciences (NISAN), Auckland University of Technology, Auckland, New Zealand
e-mail: ekta.dahiya@aut.ac.nz

A. M. Ibrahim (✉)
Fundamentals of Nursing Department, College of Nursing, Imam Abdulrahman Bin Faisal University, Dammam, Saudi Arabia
e-mail: amsudqi@iau.edu.sa

M. S. Alam
School of Medical and Allied Sciences, K. R. Mangalam University, Gurgaon, India

F. A. Khan
Department of Stem Cell Research, Institute for Research and Medical Consultations, Imam Abdulrahman Bin Faisal University, Dammam, Saudi Arabia

F. H. Pottoo (✉)
Department of Pharmacology, College of Clinical Pharmacy, Imam Abdulrahman Bin Faisal University, Dammam, Saudi Arabia
e-mail: fhpottoo@iau.edu.sa

S. Beg et al. (eds.), *Nanophytomedicine*,
https://doi.org/10.1007/978-981-15-4909-0_9

large molecular weight and poor solubility, which raises the issue of low bioavailability, thus these active constituents need to be given in large quantities which is again undesirable; the problem is largely overcome with the exploitation of nanotechological approaches, which at the same time allows flexibility in routes of administration such as transdermal, nasal, vaginal, inhalation and parental. The nanoformulations of herbal medicines formulated tend to reduce extent of hepatic first pass metabolism and gastric pH mediated drug degradation, so overall enhancing bioavailability of orally administered drugs. Further additional challenges in the form of blood brain barrier (BBB) and efflux pumps, are faced by the traditional dosage forms, containing neuroactive phytoconstituents. However, these challenges may be by large circumvented through encapsulating active phytoconstituents within nanoparticles. The trends have emerged towards judicious integration of scientific evidences related to phytoconstituents and their mechanism, selection of appropriate materials, assessment of process variables, and specific formulation approaches which includes active drug targeting using surface guided ligands to improve overall performance. These scenarios compelled pharmaceutical companies to file patents with claims on therapeutic properties of herbal medicine loaded formulations such as nanoliposomes of curcumin, solid lipid nanoparticles (SLNs) of digoxin, albumin based nanoparticles of taxane and paclitaxel as well as micellar formulations of vincristine and vinblastine.

Keywords

Herbal medicine · Flavonoids · Glycosides · Terpenoids · Phytomedicine · Phytopharmaceuticals · Nanoformulations · Drug delivery

1 Historical Preview of Progress in Phytomedicine

The World Health Organization (WHO) defined traditional medicines as a diverse health practice including the use of herbal medicines, mineral-based medicines and animal parts which are used as a single material or in combination to treat, diagnose or prevent illness and to maintain well-being. The term complementary/alternative/non-conventional medicine is used alternately with traditional medicine in some countries [1]. Several archaeological evidences proved that herbal medicines were used since prehistoric times, as pain killers, anti-inflammatories, anti-diarrhoeals, digestive aids and fertility enhancers [2]. The ancient Egyptians were the pioneers in formulating multiple curative compounds extracted from natural sources like animals, mineral and vegetable sources. Egyptians are reported to be the early users of wine, castor oil, marijuana, opium, mints and beer made from barley and wheat for medical purposes [2]. Dioscorides was famous Greek physician who authored De Materia Medica, a comprehensive book covering over 600 medicinal plants, between 50 and 70 CE and that was used for the next 1500 years and translated into several languages such as English, Arabic and Latin.

In India, the Ayurveda is considered the ancient science of Indian system of medicine and thought to be the oldest written medical system; the Ayurveda mentions more than 2000 drugs of natural sources [3]. Jabir ben Hayyan, the Arabic scientist, was one of the leading individuals who practiced medicine in the eighth century, and published a six chaptered book on Poisons and Antidotes; in the book details of natural resources of poisons, how they work, the dose and the route of administrations were mentioned [4]. The medicinal plants were also used as spices to flavour the food in cooking and to preserve food from pathogens; for example, in tropical regions where microorganisms are grown in large numbers spicy foods possess sufficient antimicrobial action [5].

The Ayurveda in India, Traditional Chinese medicine in China, Native American medicines in the USA, or Unani medicines of Southeast Asia, Rongoa/traditional Māori healing are alternative therapeutic approaches which rely on natural herbs [6]. The introduction of allopathic medicines in the western world led to a decline of herbal medicine practices across the world, but still most of it is practised in developing nations. The WHO reported that approximately 75–90% of the population of developing countries essentially depend on traditional medicinal practices [7, 8]. While, in other WHO study, 80% of the world population were reported having good faith with regular usage of traditional medicine for their primary health care as well as management of disability disorders. In countries like China, India and Tibet, herbalists are essentially valued as highly respected professionals in the society. Subsequently due to long history of safety evidences of these herbs as well as scientific validation of claims are now popularizing natural products in developed countries [9]. Herbs are popularly used as nutraceuticals or add-on therapy, and that combined dosage forms exert synergistic efficacy, thereby reducing side effects in drug regimen, in a very effective manner [10, 11]. Subsequently due to the supportive government policies, legal and regulatory frame works as well as integration of statistical quality control processes such as QbD, ease of availability, affordability, these traditional medicines are to retain an important position in the health care system [8, 12, 13].The current modern therapy and treatment approaches which involve monotherapy and sometimes polytherapy for management of complex disorders such as inflammatory, neurological disorders and cancers, provide limited and symptomatic relief [14, 15]. Hence, integration of herbs within the healthcare system may favour desired pharmacological responses [12], e.g. Aloe gel containing silver sulfadiazine (SSD) in nano-sized and nanosuspension form fetched good wound healing properties [16, 17]. It is also evident from randomized controlled trails that herbal medicines hold therapeutic benefits in terms of efficacy; however, inadequate information remains about their mechanism of actions and scientific valuation of performances of their active phytoconstituents [18, 19] (Tables 1 and 2).

2 Challenges with Phytomedicine

It is quite common that some potent and usually famous herbs from traditional systems of medicine might not be much effective in other geographical locations. The reasons may be attributed to change in geographies and climatic conditions

Table 1 Phytoconstituents and mechanism of plant based herbal therapy in Alzheimer's disease

S. No.	Plant	Active constituents	Mechanism of action	Disease	References
1	*Bacopa monniera*	Bacoside A, Bacoside, Betulinic acid, Stigmastanol, b-Sitosterol, Stigmasterol	Inhibition of cholinergic degeneration and cognition-enhancing effects	Alzheimer's Disease	[95]
2	*Centella asiatica*	Asiatic acid and asiaticoside	Inhibition of beta-amyloid formation	Alzheimer's Disease	[96]
3	*Curcuma longa*	Essential oils, Curcumin, polyphenol	Inhibition of articular NF-B in RA joints	Alzheimer's Disease	[97]
4	*Ginkgo biloba*	Ginkgolides A, B, C, J and bilobalide, biflavones, Terpene trilactones, proanthocyanidins, alkylphenols, polyprenols	Scavengers of free radicals and reducing membrane fluidity	Alzheimer's Disease	[98]
5	*Chamomilla recutita*	Bisabolol oxide A, alpha-bisabolol, bisabolol oxide B, cis-enyne-bicycloether, bisabolon oxide A, chamazulene, spathulenol and (E)-beta-farnesene	Increasing c-AMP levels, morphine dependence and withdrawal symptoms	Alzheimer's Disease	
6	*Acorus calamus*	α-and β-asarone	Inhibition of acetylcholinesterase (AChE)	Alzheimer's Disease	[99]
7	*Celastrus paniculatus*	Triacylglycerol, free fatty acids, diacylglycerol, esterified sterols and mono acylglycerol	Cholinergic activities improve memory performance	Alzheimer's Disease	[100]
8	*Commiphora mukul*	Terpenes, sesquiterpenoids, cuminic aldehyde, eugenol, and the ketone steroids Z-and E-guggulsterone, guggulsterols I, II, and III	Decreased neuronal cholesterol, AChEase activity and beta-amyloid	Alzheimer's Disease	[101]
9	*Panax ginseng*	Ginsenosides, or saponins, 20(S)-protopanaxadiol (PPD) and 20(S)-protopanaxatriol (PPT)	Inhibition of amyloid beta (Aβ) and interruption of hyperphosphorylation	Alzheimer's Disease	[102]

Table 2 Phytoconsituents and mechanism of plant based herbal therapy in Parkinson's disease

S. No.	Plant	Active constituents	Mechanism of action	Disease	References
1	*Allium sativum*	S-allylcysteine	Improved locomotor, superoxide dismutase (SOD) activity, and dopamine (DA) levels	Parkinson's Disease	[103]
2	Cameliasinesis	Epicatechin-3-gallate	Attenuate DA depletio, inhibit DA uptake by blocking	Parkinson's Disease	[104]
3	*Cassia obtusifolia*	Rubrofusarin, Isorubrofusarin	Attenuate reactive oxygen species (ROS) generation and mitochondrial membrane depolarization	Parkinson's Disease	[99]
4	*Coffea arabica*	Caffeine	Reversal of motor deficit and apomorphine-induce rotation	Parkinson's Disease	[105]
5	*Lycopersicon esculentum*	Lycopene	Decreased DA loss	Parkinson's Disease	[106]
6	*Nigella sativa*	Thymoquinone	Protect dopaminergic neuronal deprivation	Parkinson's Disease	[107]
7	*Sesamum indicum*	Sesamin, sesamol, sesaminol	Modulates tyrosine hydroxylase, SOD, and catalase levels.	Parkinson's Disease	[108]
8	*Uncaria rhynchophylla*	Rhynchophylline, corynoxeine, corynantheine, hirsutine	Reduction in ROS and caspase 3 activity	Parkinson's Disease	[99]
9	*Withania somnifera*	Withanone, withaferin, withanolides, withasomidienone, withanolide	Inhibit metastasis and quinone reductase activity	Parkinson's Disease	[109]
10	*Zingiber officinale*	Zingerone and 6 shogaol	Impede 6-OHD, induced DA loss and prevent neuronal apoptosis	Parkinson's Disease	[110]

which affect the overall phytoconstituent levels, such as difference of polyphenol contents reported in *Terminalia chebula*, *Terminalia bellerica* and *Emblica officinalis*. So, it is quite obvious that these herbs would exhibit different performance at clinical stage. Adding to the challenge, even herbal medicines collected from same sources sometimes failed to exert same responses; reasons could be lack of consideration on pharmacogenomics aspects of these herbs [20].

Furthermore, the plants are usually rich in many therapeutically active ingredients so in order to predict in vivo performances, screening of individual phytoconstituent as well as its quantification, further complicates clinical translation, in effective manner. Due to lack of scientific endorsement on pharmacological response of herbal drugs as well as *in vivo* mechanisms, prediction of herb–allopathic drug interactions makes it more complicated, while such prediction of performances are more complicated in various neuronal diseases and cancers which inherently exhibit cross-talk mechanisms [21].

Apart from above, herbs and biological macromolecules, such as proteins, also show limited performance in clinical scale through their conventional dosage forms because of various contributing factors and pharmacotechnical challenges. These herbal phytoconstituents contain aromatic hydrocarbon rings in their structures which contributes towards poor aqueous solubility. Such drugs, for example bioactive Curcumin (isolated from *Curcuma longa*), exhibit poor aqueous solubility, and are released at lesser extent and with slow rate in stomach and intestinal milieu. This dissolution based pharmacotechnical challenge of Curcumin, which reduces overall bioavailability of drug, is defined as rate limiting step in conventional drug delivery system of curcumin [22]. However, there are various formulation techniques available to enhance solubility of poorly aqueous soluble drugs in order to achieve higher overall bioavailability. Curcumin and polyphenols are more stable in stomach but they exhibit low aqueous solubility. While some drugs such as Mitragynine (psychoactive alkaloid) are more liable to undergo high degradation at stomach (acidic) pH [23]. Hence, for such acidic liable drugs many attempts are being adopted to not only ensure lesser extent of drug release in stomach, but also higher extent of loaded drug release, once it reaches intestinal region (alkaline pH). Preferably by adopting prodrug concepts or by integrating pH responsive fabricating materials in dosage forms [24]. Subsequently, lipophilic herbal drugs preferably undergo pre-systemic liver metabolism, before reaching the target site of action. While, many of times, fraction of drugs which undergo biotransformation in the liver, were not only reported to exhibit reduced potency and efficacy but also few of their analogues may either be ineffective or sometimes toxic as well. For example, many of pyrrolizidine class of alkaloids which are present as secondary metabolites of some flowering plants are toxicologically inert from in vitro testing. However, in in vivo model, the post metabolic activation derived congeners lead into severe consequences such as hepatotoxicity, mutagenicity, and carcinogenicity [25]. In lieu of above arguments, prima facie goal of drug delivery i.e. is to ensure the release of active constituents from formulation, at the site of action, in well controlled, safe and predictable manners, which are difficult to attain with traditional drug delivery techniques. Thus to overcome these above mentioned limitations in herbal practice,

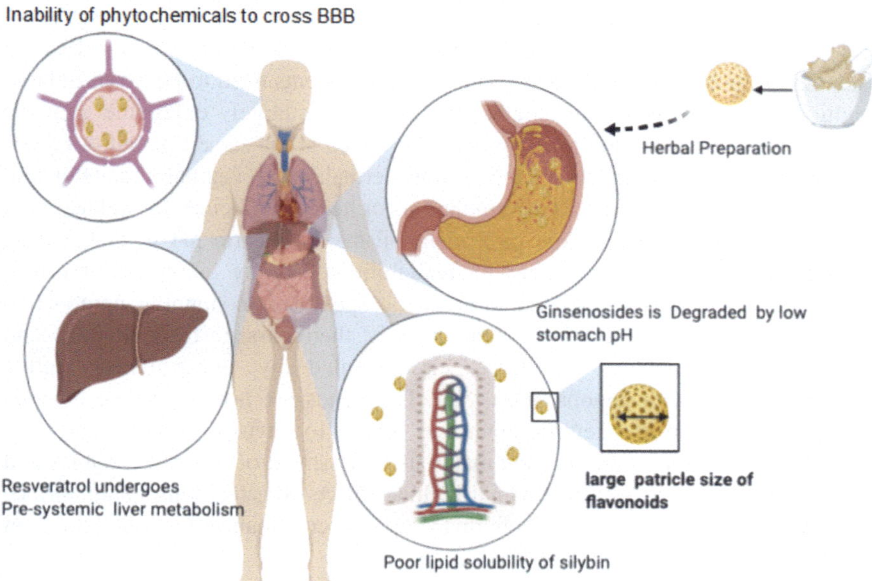

Fig. 1 Challenges with phytomedicine delivery

judicious formulation approaches and advanced technological skills are to be availed to fabricate design and development of novel drug delivery systems [26, 27] (Fig. 1).

3 Nanotechnological Advancements in Phytomedicine Based Therapy

The advances in the modern technological based extraction, isolation, and identification of active constituents from therapeutically active herbs validate their traditional therapeutic benefits, claimed from a long time. Furthermore, current approaches for co-delivery of these pharmacologically active components with drugs from diverse drug categories like anti-diabetic, neuromodulators, hypolipidemic agents, anti-infectious modulate overall therapeutics responses, especially for diseases in which complex interlinked pathophysiological pathways are reported. Thereby in such cases, the unique mechanism of action of these herbal drugs, when used as adjuvant therapies significantly improves action of existing allopathic drugs in diverse disease states. Traditionally most of these herbal medicines were taken either in form of powders, teas or extracts. However, nowadays these are usually taken as dietary supplements preferably in form of tablets, capsules or ointments [28]. In the past, juices, paste and decoction of these herbs were exploited as drug delivery systems for most of the herbal preparations but due to low shelf life of these preparations, nowadays herbs are formulated in form of fermented syrups, and medicated oils. The active constituents of herbal drugs such as, flavonoids, glycosides, terpenoids, are poorly absorbed from the gastrointestinal tract because of their large molecular weight and poor solubility which raises the issue of low bioavailability; thus these active constituents need to be given in large quantities which is again undesirable; the problem is largely overcome with the exploitation of nanotechnological approaches, which at the same time allows flexibility in routes of administration such as, transdermal, nasal, vaginal, inhalation and parental routes [29, 30]. The nanoformulations of herbal medicines formulated tend to reduce extent of hepatic first pass metabolism and gastric pH mediated drug degradation, so overall enhancing bioavailability of orally administered drugs. Further the challenges faced by traditional dosage forms containing neuroactive phytoconstituents faces additional challenges of their passage through BBB and efflux pumps which is by large circumvented by encapsulating active phytoconstituents with nanoparticles [31]. The trends are now emerging towards judicious integration of scientific evidences related to phytoconstituents and their mechanism, selection of appropriate materials, assessment of process variables and specific formulation approaches which include active drug targeting using surface guided ligands, to improve overall pharmacological performances of loaded drug, on both preclinical and clinical stages [32].

3.1 Rationale of Nanotechnology Enabled Delivery of Phytomedicine

As far as formulation approaches are concerned, traditional system of medicines is technologically well behind the modern system of medicines; the processing of raw materials and dosage forms in traditional systems are quite different. Here, Khamira (semi-solid oral dosage forms) and arista syrups are unique type of traditional fermented products prepared from raw plant based materials while, bhasmas are unique in sense of their mineralo-herbal origin and processing [33].

While, nanotechnology is a science that formulates the dosage forms in the size range of 1–100 nm at atomic and molecular level using safe, natural and synthetic biodegradable polymers, lipids and polysaccharides [34, 35]. These small sized nanoparticle carriers are utilized to deliver pharmacological active constituents, sourced either from allopathic or traditional system of medicines. In nanoparticulated drug delivery system, drugs are either embedded in a matrix system or encapsulated or sometimes may be adsorbed on surface. For example, nanospheres are highly porous in nature and loaded drugs are uniformly dispersed within matrix. Whereas, nanocapsules are of reservoir type in which core region is occupied with drugs [36–38]. These nanoparticulated systems may be further differentiated into various ways based on fabricating materials, such as metallic nanoparticles which contain anti-cancer nobel metals such as gold (Au) and Silver (Ag) or anti-microbial oxide forms of metals such as oxide of Zinc (ZnO) or Magnetic oxide of Copper (CuO); Polymeric systems in forms of film, patches, drug reservoirs and gels; lipid based nanoparticles such as emulsion, SLNs, nanolipid carriers (NLC), cubosomes, limicubes, liposomes, etc. [39]. Usually materials used to fabricate these systems inherently bear either the negative or neutral charges, except few such as chitosan. While, as a rule of thumb; lower the charge over particles (zeta potential), lower would be the stability of such fabricated systems [40]. The negative charge of the surface and nanoparticles may repel each other, so limiting enhanced permeation and retention (EPR) effects. Hence, some cationic polymers such as chitosan and lipids such as 2,3-dioleoyloxy-propyl-trimethylammonium chloride are also employed which show better interaction of their nanoparticulated systems with the pathogenic cells [41].

In brief, nanofabrication technology with nanoparticles enables target drug delivery at specific sites, improving pharmacokinetic profile and pharmacodynamic performance and thus reduce dosing frequency, coupled with reduction in side effects due to dose reduction. Further advantages include favoured dissolution improved EPR of loaded drugs at the site of action [42]. The systems which bear EPR impact will obviously exhibit higher permeation through the biological barriers which acted as rate limiting steps to limit fraction of drug exposure at site of drug action. So, most fractions of loaded drugs, due to higher permeation attributes of such carriers, ensure to reach within the cells, to maximum extent even at low doses. Many of herbal drugs which are not being absorbed topically due to presence of skin as barrier are being solubilized in oil for higher absorption of drugs loaded oil droplets [43]. While, other important aspect which affect therapeutic response of administered drug is retention impact of drug. Pathogenic cells are bearing many efflux pumps and some other types of defensive mechanisms, which by identifying these drug molecules as foreign objects expell out the absorbed fraction of herbal drugs. Hence, higher the retention impacts of drug, higher will be their therapeutic impacts and longer will be duration of actions. These EPR impacts are very important consideration in neurological diseases and cancers because of highly defensive barriers and efflux pump activation in the central nervous system (CNS) [42].

Although, very recently, we shifted our focus on nanotechnology to deliver the drugs but many of ancient and traditional processing steps were being used with herbal drugs in order to potentiate their action. Among these, metallic Bhasma, a type of prepration, involved steps in the formulation leading for the growth of particles in the nanoscale ranges. However, they were not concerned for their dimension aspects but evidences suggest their significant potentiation in clinical settings, they were quite sure that these processing steps integrate some unique attributes into drug which favour overall performances of drugs in various diseases. Hence, adaption of these approaches with herbal medicines or phytoconstituents enable improvement in overall biopharmaceutical performances of loaded drugs [44]. The judicious integration of nanotechnology based approaches with either crude drugs or phytoconstituents of hydrophilic and/or hydrophobic components offers better solutions in terms of site specific targeted drug delivery. Banking on prospects and generation of interests, for integration of nanotechnology based concepts with herbal drugs, translated into filling of numerous patents, across the world. The scientific validation of therapeutic promises of phytoconstituents, calls for adaptation of more stringent and standardized regulatory enforcement with herbal drugs and their nanotechnology based formulations, as motivational and guiding force, to continue maintaining trust of healthcare practitioners and patients, in these modernized drug delivery systems.

4 Drug Delivery Systems for Phytochemicals

As a drug delivery system different types of nanoparticles are used to deliver phytochemicals, namely polymeric nanoparticles, nanospheres, nanocapsules, SLNs, liposomes, micelles, liquid crystals, and dendrimers [37, 38] (Fig. 2; Table 3).

4.1 Liposomes

Liposomes are bipolar compounds that are having lipophilic and hydrophilic moieties at the same time; phospholipids are the main structural units of these vesicles. The spherical shape encapsulates the drug inside it, which floats freely inside the vesicles [45, 46]. Once in the water, the hydrophilic part orients itself toward the water, while the lipophilic part faces the inner part away from the aqueous region forming a bilayer membrane. Consequently, both hydrophobic and lipophilic materials can be encapsulated inside the liposomes. Liposomal preparations have

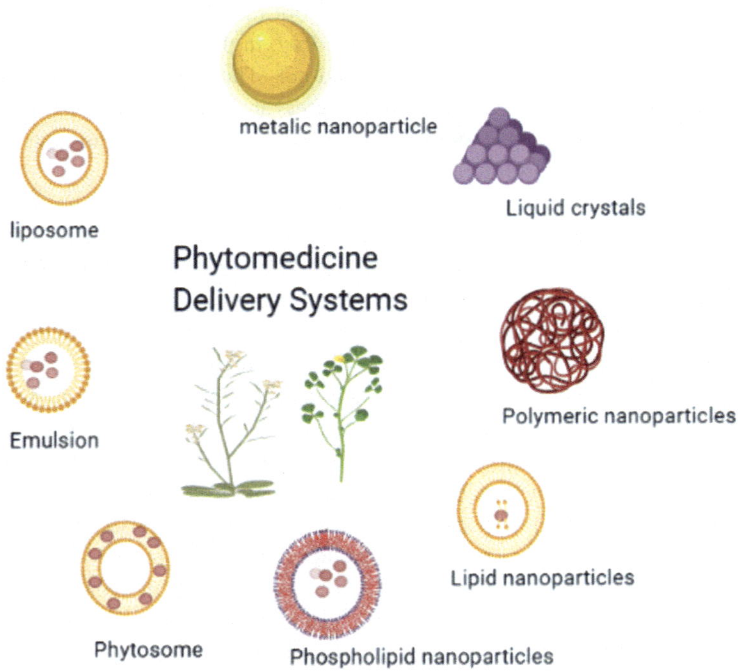

Fig. 2 Drug delivery systems for phytochemicals

Table 3 Biopharmaceutical aspects of nanotechnology processed formulation of Phytoconstituents

S.No.	Phytoconstituents	Systems	Materials	Performance			References
				Parameters	Drug	Formulation	
1	Bacoside-A	Polymeric nanoparticles	PLGA	Brain concentration (μg/g tissue)	2.56	23.94	[111]
2	Asiatic acid	Albumin nanoparticles	Albumin	Brain AUC (h. ng/g)	17,306	541,060	[112]
3	Curcumin	Polymeric micelles	Polyethylene glycol–Polylactide (PEG-PLA)	Plasma AUC (min × μg/g)	0.1 ± 0.1	0.5 ± 0.2	[113]
4	Ginkgolides A and B	Nanosuspension	Soybean phospholipid	Relative bioavailability (%)	A: 100 B: 100	A: 221 B: 437	[114]
5	α-bisabolol	SLNs	Cholesterol	AChE inhibitory and Anti-Amyloidogenic property	Insignificant protection	Significant protection	[115]
6	α-asarone	Nanoparticles	mPEG-PLA	Q_{18h} (μg/cm2)	223	382	[116]
7	Celastrol	Self-assembled phytosomalnanocarriers	Phospholipid	AUC 0– t ng/mL.h	767	186	[117]
8	Ginsenoside Rg3	Liposomes	Phosphatidylcholine	AUC(0–t) (g/l h)	382	582	[118]

improved solubility, increased bioavailability and enhanced uptake by the cells. This will increase the therapeutic activity and safety profile of the herbal preparations or phytochemicals [47]. An obvious example for liposomal herbal preparations is Orthosiphon stamineus extract nanoliposomes; with improved bioavailability and brain penetration, this preparation could be used in the treatment of Parkinson's disease as it has in vitro antioxidant activity. Curcumin encapsulated nanoliposomes in animal testing appeared to have batter bioavailability after oral administration with improved antioxidant effect. Curcumin in nanoliposomes possesses the tendency to cross the blood–brain barrier and are reported effective against Alzheimer's disease [48]. Dihydroartemisinin nanoliposomes (DHA-MLPs) have significant antitumor efficiency and bioavailability, and could be used for the treatment of head and neck tumours [49].

4.2 Phytosome

Phytosome is the technology in which standardized herbal extract or water-soluble phytoconstituents are bound to phospholipid compound to enhance the lipid solubility and to increase drug delivery. Compared to liposomes, in which there is no chemical bond between the phosphatidylcholine and the water-soluble substance, phytosomes possess bonds (stoichiometric ratio—1:1 or 1:2) between phosphatidylcholine and water-soluble substance. Many of the biologically active compounds from herbal sources are usually water-soluble constituents, like flavonoids, tannins, terpenoids. Which are thus unable to cross the lipid membrane, resulting into minimal bioavailability and reduced therapeutic effects. To overcome solubility issues and to increase pharmacological effects of the herbal extracts, these carriers bind with phosphatidylcholine form the phytosome, to exhibit better bioavailability with respect to conventional herbal extracts, owing to improvement in lipophilicity [50]. Recently, phytosomes become popular delivery system for herbal and nutritional compounds. Ginkgoselect® phytosome is a herbal preparation of *Ginkgo biloba* leaves, compared to the standardized extract it showed enhanced absorption rate. It is mainly used for cases of cerebral insufficiency and peripheral vascular disorders, as it improves cerebral blood flow. Beside its improved bioavailability, it has also good tolerability which makes it good for longer period of treatment [51].

4.3 Polymeric Nanoparticles

Polymeric nanoparticles are derived from natural (polysaccharides—chitosan, cellulose; proteins—gelatin, albumin) and/or synthetic [polyethylene glycol (PEG), polyvinyl alcohol (PVA), polylactic acid (PAA)], polylactic glycolic acid (PLGA) sources. Examples of herbal drugs formulated with polymeric nanoparticles include root extract of *Phytolacca decandra* with PLGA-encapsulation for lung cancer [52], leaf extract of *Ocimum sanctum* as anti-microbial [53], rhizome extract of *Curcuma longa* encapsulated in PLGA nanospheres with anti-tumour properties [54]. Recently cellulose acetate electrospun nanofibres have been used for natural biodegradable polymer delivery in transdermal or wound dressing patches [55].

4.4 Phospholipid Nanoparticles

Phospholipids have natural origin, act as a carrier and could be of two types, liposomes or micelles. The only difference is bi- or single layer of phospholipid. Liposomes are composed of bi-lipid layer suitable for both hydrophilic and hydrophobic environment, while micelle have single lipid layer facing toward the core favouring the hydrophilic environment like blood [37, 38, 56]. A Cratylia-molislectin loaded liposomes have shown immunostimulatory action in Swiss mice [57]. While, suspension of Quercetin-liposomes has shown anxiolytic and cognitive activities in Wistar rats [58]. Similarly, enhanced antioxidant properties of α-tocopherol studied *in vitro* with green tea polyphenols solution in sodium dodecyl sulphate [59], are clinically significant application of phospholipids formulations in delivery systems.

4.5 Emulsions

Emulsion is the mixture of two layers of fluids (usually oil and water) which under normal conditions does not form homogenous phase; but during the emulsification processes, one of the liquid start to disperse in other phase to form droplet particles. These systems are broadly classified into 3 different types according to the particle diameter: regular emulsion (0.1–100 μm), micro and nanoemulsion (10–100 nm) and sub-micro (100–600 nm) also called lipid emulsion. Perfect quantity of specific surfactant in fixed ratio of oil water make them suitable for administration via different routes. These emulsifying systems can be used for oral and injectable dosage forms. Emulsions have higher affinity to the lymph that makes them to be distributed in a targeted way and thereby can be used for intramuscular or subcutaneous injections, and as sustained release formulations. Herbal phytoconstituents formulated as emulsions are more stable with improved ability to penetrate the skin and mucous membranes [60]. Vitamin E nanoemulsion was formulated using the spontaneous emulsification technique which reduced neuronal oxidative stress in Parkinson's disease. A pomegranate seed oil nanoemulsion reduced neuronal loss with strong protective effects [48]. Curcumin (Cur) nanoemulsion (NE) played a significant role in wound healing and exhibited anti-inflammatory effects [61].

4.6 Lipid Nanoparticles

Solid lipid nanoparticles are 1st generation solid nanoparticles of lipid based targeted delivery system and are considered improved version of phospholipid nanoparticle systems. The hydrophobic drugs are being incorporated in association with some stabilizing agents to obtained homogenous matrix of solid in a carrier solution. While second generation of lipid based drug delivery systems, are called nanolipid carriers (NLC), in which some liquid lipids are also incorporated, in association with solid lipid, to improve overall stability of loaded drugs and to obtain desired release performances. NLC exhibits core-shell type of model; drugs are usually present in the core but sometimes in shell as well. The nanostructure lipid carriers are now used to disperse the solid drug in an imperfect matrix, structure less solid amorphous matrix or multiple carrier with multiple oil in fat in water [37, 38, 62]. Quercetin-SLN was formed using the emulsification-solidification method which increased bioavailability by five times, making it more soluble [63]. Paclitaxel-NLCs for treatment of the liver cancer are reported with improved activity [64].

4.7 Liquid Crystals

State of matter of liquid crystals is in between crystalline solid and an isotropic liquid. They are composed of anisotropic molecules which in response to external fields modify their structure and properties by interacting on surface. Liquid crystals are categorized as hydrophobic anisometric type with thermal phase, hydrophilic type with structural phase, and amphotropic including both hydrophobic and hydrophilic type showing thermal and structural phases [38, 65, 66]. Improved physical stability was achieved with *Prunus persica* essential oil by formulating it into liquid crystal which was oil-in-water type of emulsion systems [67]. Using the hydrophilic/lipophilic balance method a stable liquid crystal was formed of marigold oil [68]. Limicubes which are cuboidal shaped lipid crystalline particles of natural occurring lipid Monoolein, promised improved delivery of loaded drug rosuvastatin and enhanced pharmacological performances [27].

4.8 Metallic Nanoparticles

New techniques of formulating nanoparticles are being researched, such as use of metallic nanoparticles which include silver, gold, and magnetic nanoparticles [35]. Concept and integration of these metallic nanoparticles are not new rather many traditional systems of medicines such as Unani and Ayurveda discussed several process of making traditional formulations which includes but not limited to Bhasma, to yield range of particles in nano-dimension range for enhanced therapeutic advantages. Although there are several chemical based nanoparticle synthesis approaches but recent trends are towards adaption of green chemistry approaches using polymers or plant extracts to fabricate these nanoparticles e.g. gum ghatti was used in the synthesis of stable gold nanoparticles [69] and extract from purple heart plant leaves for the synthesis of silver nanoparticles; the latter possesses significant anti-microbial property against clinically pathogenic strains of microbes [70–72].

5 FDA Approved Drugs

Some of the FDA approved nano-based herbal drugs are albumin bound paclitaxel (Abraxane, Celgene) for lung cancer, primary and secondary metastatic breast cancer [73], liposomal non-PEGylated vincristine (Marqibo, Spectrum) for acute lymphoblastic leukaemia, liposomal PEGylated Irinotecan—a semisynthetic derivative of Camptothecin for metastatic pancreatic cancer, and nanocrystal of Morphine sulphate (Avinza, Pfizer) as psychostimulant [74, 75]. Collectively all these approved nano-pharmaceuticals have shown improved efficacy, reduced toxicity when compared with their conventional herbal active constituent. There are several on-going clinical trials of herbal drugs. The approved anti-cancer drugs such as Doxil (doxorubicin), Marqibo (vincristine) and Onivyde (irinotecan) are undergoing trials for additional cancer types, combination therapy, and upgrading to primary first-line therapies. On the same line, FDA approved drug AmBisome is under trails for additional anti-bacterial and anti-fungal properties [74]. The advances in the field of nanotechnology are paving the way for the clinical use of herbal medicines.

6 Recent Patents

The aim of any patented research is to get the invention to the masses. It has been observed that in the last 30 years, out of all approved natural compounds by the US Food and Drug Administration and other international regulatory bodies nearly 61% are for treating cancer, and the rest 49% are for treating infection [38, 76]. Most number of patents with therapeutic properties are nanoparticles of curcumin using liposomes, SLNs, polylactic-co-glycolic acid as carriers [77]. Patents of plant extracts approved for anti-cancer properties include water-soluble extract of *Solanum* genus and Curcumin [78, 79], liposomal formulation of Digoxin, Digitoxin [80] and Curcumin [81] nanoparticle of taxane and paclitaxel with albumin as carrier [82, 83], micellar formulation of Vincristine and Vinblastine [84], liposome-encapsulated complex of terpenes [85]. Different nanoparticulate formulations such as SLNs and microemulsion of Curcumin have been patented for treating Alzheimer's disease [83, 86]. In order to cross BBB, chloroquine nanoparticles were developed to increase the cellular respiration, for the clinical management of Parkinson's disease [77, 87]. A patent filed by Aburdeinch et al. (2012) claimed to reduce cholesterol levels by 30% with oral administration of fenugreek seeds extract through nanoparticle carriers for 30 days [88]. A progressive bone disease, osteoporosis, was treated with a formulation of *Butea monosperma* extract as nanoparticle and nanoemulsion [89]. Another well explored area of patented nano-herbal formulations is cosmetic or topical application. A controlled release or an ion-paired nano-diffusion system of number of plant extract using 1–70% of zeolite as a carrier presented with activities such as promoting cellulite control, excess fat reduction, toning, collagen synthesis, increasing skin elasticity, skin whitening, anti-microbial, protecting from ultraviolet radiations and anti-oxidant properties [90, 91]. Nanocapsule or microemulsion of capsaicin for acute, chronic or neuropathic pain was patented by Bakthavatchatam and group [92]. Extracts of citrus, cinnamon, tocopherol, carotenoid, rosemary, thymol, menthol and eucalyptol when dispersed in nanoparticle solution exhibited desired effects in oral health care [87, 93, 94].

7 Conclusion and Future Prospective

The term nanomedicine is often used to describe drug loaded novel carrier systems which are being fabricated through nanotechnology based processes for the purpose of their effective exploitation either in the treatment, diagnosis or prevention from diseases. Recently, there are growing concerns for judicious integration of concept and approaches of nanotechnology with phytomedicines. Successes and prospects of such formulation at clinical level owing to enhancement of absorption and bioavailability, rapid dissolution, quicker onset of action, higher plasma concentration, prolonged actions, site specific targeting, reduction in side effects and toxicity as well as improvement in pharmacological performances of loaded phytomedicine; already generated a lot of interests and trends among researchers to exploit them in the treatment and control of several chronic diseases such as: cancer, arthritis, neurological disease such as, Parkinson's, Alzheimer's, and epilepsy. In brief, integrating advanced technology in processing herbal medicine based preparations attract pharmaceutical industries to provide better alternatives than conventional dosage forms, in order to improve the health of people through these "herbal remedies". Despite tremendous prospects and advantages of such nanotechnology based nanophytomedicinal meagre products found their ways in the clinical practices; owing to involvement of complex and advanced machinery processes, scale-up issues, high cost of preparation, environmental and health threats from generated tiny sized particles, low shelf life of such developed formulations; which are still considered as major barriers, hence need to be addressed, in very systemized way.

References

1. Alghamdi M (2018) Herbal medicine use by Saudi patients with chronic diseases: a cross-sectional study (experience from Southern Region of Saudi Arabia). J Health Special 6(2):77
2. Halberstein RA (2005) Medicinal plants: historical and cross-cultural usage patterns. Ann Epidemiol 15(9):686–699
3. Mukherjee PK (2001) Evaluation of Indian traditional medicine. Drug Inf J 35(2):623–632
4. Saad B (2006) Safety of traditional Arab herbal medicine. Evid Based Complement Alternat Med 3(4):433–439
5. Seema A (2014) Recent development of herbal formulation—a novel drug delivery system. Int Ayurved Med J 2(6):952–958
6. Chatfield K, Salehi B, Sharifi-Rad J, Afshar L (2018) Applying an ethical framework to herbal medicine. Evid Based Complement Alternat Med [Internet]. [cited 2019 Nov 6];2018. Available from: https://www.ncbi.nlm.nih.gov/pmc/articles/PMC6169226/

7. Bonifácio BV, da Silva PB, Ramos MADS, Negri KMS, Bauab TM, Chorilli M (2014) Nanotechnology-based drug delivery systems and herbal medicines: a review. Int J Nanomedicine 9:1–15
8. Zhang X (2004) Traditional medicine: its importance and protection. Protecting and promoting traditional knowledge: systems. Natl Exp int Dimens Part 1:3–6
9. Mishra S (2019) Bioinspired nanocomposites: applications in disease diagnosis and treatment. Pharm Nanotechnol 7(3):206–219
10. Nigar S (2016) Molecular insights into the role of inflammation and oxidative stress in epilepsy. J Adv Med Pharm Sci 10(1):1–9
11. Pottoo FH (2019 Feb) The synergistic effect of raloxifene, fluoxetine, and bromocriptine protects against Pilocarpine-induced status Epilepticus and temporal lobe epilepsy. Mol Neurobiol 56(2):1233–1247
12. Ahmad N (2019) Tamarindus indica fruit: Pharmacognostical standardization, detection of contaminant, and in vitro antioxidant activity. J Pharm Bioallied Sci 11(4):355
13. Javed MN (2019) QbD applications for the development of Nanopharmaceutical products. Pharm Qual Design 24:229–253
14. Patra P (2018) Phytochemical investigation and hepatoprotective effect of Scoparia dulcis against carbon tetrachloride induced liver damage in rats. Int J Pharm Sci Res 9(7):1086–1092
15. Pottoo FH (2019) Estrogen and serotonin: complexity of interactions and implications for epileptic seizures and Epileptogenesis. Curr Neuropharmacol 17(3):214–231
16. Barkat MA (2017) Nanosuspension-based Aloe vera gel of silver sulfadiazine with improved wound healing activity. AAPS PharmSciTech 18(8):3274–3285
17. Barkat MA (2018) Therapeutic intervention of gel containing Nano-sized and Micron-sized silver sulfadiazine gel on second-degree burn: a comparative study. Int J Low Extrem Wounds 17(3):176–183
18. Jachak SM, Saklani A (2007) Challenges and opportunities in drug discovery from plants. Curr Sci 92(9):1251–1257
19. Parasuraman S (2018) Herbal drug discovery: challenges and perspectives [Internet]. Curr Pharmacogenomics Personalized Med. [cited 2019 Nov 6]. Available from: http://www.eurekaselect.com/161425/article
20. Palav YK, D'mello PM (2006) Standardization of selected Indian medicinal herbal raw materials containing polyphenols as major phytoconstituents. Indian J Pharm Sci [Internet] 68(4). [cited 2020 Feb 27]. Available from: https://www.ijpsonline.com/abstract/standardization-of-selected-indian-medicinal-herbal-raw-materials-containing-polyphenols-as-major-phytoconstituents-3201.html
21. Pottoo FH, Tabassum N, Javed MN, Nigar S, Sharma S, Barkat MA, Harshita, Alam MS, Ansari MA, Barreto GE, Ashraf GM (2020 Feb 13) Raloxifene potentiates the effect of fluoxetine against maximal electroshock induced seizures in mice. Eur J Pharm Sci 146:105261
22. Mohammadian M, Salami M, Momen S, Alavi F, Emam-Djomeh Z, Moosavi-Movahedi AA (2019) Enhancing the aqueous solubility of curcumin at acidic condition through the complexation with whey protein nanofibrils. Food Hydrocoll 87:902–914
23. Basiliere S, Kerrigan S (2020) Temperature and pH-dependent stability of mitragyna alkaloids. J Anal Toxicol [Internet]. [cited 2020 Feb 27]; Available from: https://academic.oup.com/jat/advance-article/doi/10.1093/jat/bkz103/5693674
24. Ghaffar A, Yameen B, Latif M, Malik MI (2020) Chapter 14 - pH-sensitive drug delivery systems. In: Shah MR, Imran M, Ullah S (eds) Metal nanoparticles for drug delivery and diagnostic applications [Internet]. Elsevier, pp 259–278. [cited 2020 Feb 27] (Micro and Nano Technologies) Available from: http://www.sciencedirect.com/science/article/pii/B9780128169605000148
25. Louisse J, Rijkers D, Stoopen G, Holleboom WJ, Delagrange M, Molthof E, Mulder PPJ, Hoogenboom RLAP, Audebert M, Peijnenburg AACM (2019) Determination of genotoxic

potencies of pyrrolizidine alkaloids in HepaRG cells using the γH2AX assay. Food Chem Toxicol 131:110532
26. Dhiman A, Nanda A, Ahmad S (2012) Novel herbal drug delivery system (NHDDS): the need of hour. IPCBE 49:171–175
27. Javed MN, Kohli K, Amin S (2018) Risk assessment integrated QbD approach for development of optimized Bicontinuous Mucoadhesive Limicubes for oral delivery of Rosuvastatin. AAPS PharmSciTech 19(3):1377–1391
28. Choudhary A (2011) An overview of advances in the standardization of herbal drugs. J Pharm Educ Res 2(2):55
29. Devi VK, Jain N, Valli KS (2010) Importance of novel drug delivery systems in herbal medicines. Pharmacogn Rev 4(7):27–31
30. Haneef J, Shaharyar M, Husain A, Rashid M, Mishra R, Siddique NA, Pal M (2013) Analytical methods for the detection of undeclared synthetic drugs in traditional herbal medicines as adulterants. Drug Test Anal 5(8):607–613
31. Khan FA, Almohazey D, Alomari M, Almofty SA (2018) Impact of nanoparticles on neuron biology: current research trends. Int J Nanomedicine 13:2767–2776
32. Pottoo FH, Barkat MA, Harshita AMA, Javed MN, Sajid Jamal QM, Kamal MA (2019) Nanotechnological based miRNA intervention in the therapeutic management of neuroblastoma. Semin Cancer Biol [Internet]. [cited 2019 Oct 26]; Available from: http://www.sciencedirect.com/science/article/pii/S1044579X1930224X
33. Yousuf S, Salim S, Ahmad M, Ahmed AS, Ansari MA, Islam F (2005) Protective effect of Khamira Abresham Uood Mastagiwala against free radical induced damage in focal cerebral ischemia. J Ethnopharmacol 99(2):179–184
34. Ansari SH, Islam F, Sameem M (2012) Influence of nanotechnology on herbal drugs: a review. J Adv Pharm Technol Res 3(3):142–146
35. Chakraborty K, Shivakumar A, Ramachandran S (2016) Nano-technology in herbal medicines: a review. Int J Herb Med 4:21–27
36. Leonarduzzi G, Testa G, Sottero B, Gamba P, Poli G (2010) Design and development of Nanovehicle-based delivery Systems for Preventive or therapeutic supplementation with flavonoids. Curr Med Chem 17(1):74–95
37. Verma H, Prasad SB, Yashwant SH (2013) Herbal drug delivery system: a modern era prospective. Int J Curr Pharm Rev Res 4:88–101
38. Watkins R (2015) Natural product-based nanomedicine: recent advances and issues. Int J Nanomedicine 10:6055–6074
39. Sharma S, Noushad Javed M, Pottoo F, Arman Rabbani S, Barkat MA, Harshita, Sarafroz D, Amir M (2019) Bioresponse inspired nanomaterials for targeted drug and gene delivery. Pharm Nanotechnol 7(3):220–233
40. Alam MS, Javed MN, Pottoo FH, Waziri A, Almalki FA, Hasnain MS, Garg A, Saifullah MK (2019) QbD approached comparison of reaction mechanism in microwave synthesized gold nanoparticles and their superior catalytic role against hazardous nitro-dye. Appl Organomet Chem 33(9):e5071
41. Wang R, Zhao Z, Han Y, Hu S, Opoku-Damoah Y, Zhou J, Yin L, Ding Y (2017) Natural particulates inspired specific-targeted codelivery of siRNA and paclitaxel for collaborative antitumor therapy. Mol Pharm 14(9):2999–3012
42. Tahara Y, Yoshikawa T, Sato H, Mori Y, Zahangir MH, Kishimura A, Mori T, Katayama Y (2017) Encapsulation of a nitric oxide donor into a liposome to boost the enhanced permeation and retention (EPR) effect. Med Chem Commun 8(2):415–421
43. Sintov AC, Botner S (2006) Transdermal drug delivery using microemulsion and aqueous systems: influence of skin storage conditions on the in vitro permeability of diclofenac from aqueous vehicle systems. Int J Pharm 311(1–2):55–62
44. Bhowmick TK, Suresh AK, Kane SG, Joshi AC, Bellare JR (2009) Physicochemical characterization of an Indian traditional medicine, Jasada Bhasma: detection of nanoparticles containing non-stoichiometric zinc oxide. J Nanopart Res 11(3):655–664

45. Lasic D (1998) Novel applications of liposomes. Trends in Biotechnol (Regular ed) 16 (7):307–321
46. Kumar B, Pandey M, Pottoo FH, Fayaz F, Sharma A, Sahoo PK (2020) Liposomes: novel drug delivery approach for targeting Parkinson's disease. Curr Pharm Des. https://doi.org/10.2174/1381612826666200128145124
47. Nagalingam A (2017) Drug delivery aspects of herbal Medicines. In: Japanese Kampo medicines for the treatment of common diseases: focus on inflammation. Academic Press, London, p 143
48. Ganesan P (2015) Recent trends in the development of nanophytobioactive compounds and delivery systems for their possible role in reducing oxidative stress in Parkinson's disease models. Int J Nanomedicine 10:6757–6772
49. Li H (2019) Effects of magnetic dihydroartemisinin nano-liposome in inhibiting the proliferation of head and neck squamous cell carcinomas. Phytomedicine (Stuttgart) 56:215–228
50. Manach C (2004) Polyphenols: food sources and bioavailability. Am J Clin Nutr 79 (5):727–747
51. Shivanand P (2010) Phytosomes: technical revolution in phytomedicine. Int J Pharmtech Res 2 (1):627–631
52. Das J (2012) Poly (lactide-co-glycolide) encapsulated extract of Phytolacca decandra demonstrates better intervention against induced lung adenocarcinoma in mice and on A549 cells. Eur J Pharm Sci 47(2):313–324
53. Rajendran R, Radhai R, Kotresh TM, Csiszar E (2013) Development of antimicrobial cotton fabrics using herb loaded nanoparticles. Carbohydr Polym 91(2):613–617
54. Mukerjee A, Vishwanatha JK (2009) Formulation, characterization and evaluation of curcumin-loaded PLGA nanospheres for cancer therapy. Anticancer Res 29(10):3867–3875
55. Khoshnevisan K, Maleki H, Samadian H, Shahsavari S, Sarrafzadeh MH, Larijani B, Dorkoosh FA, Haghpanah V, Khorramizadeh MR (2018) Cellulose acetate electrospun nanofibers for drug delivery systems: applications and recent advances. Carbohydr Polym 198:131–141
56. Semalty A (2010) Supramolecular phospholipids–polyphenolics interactions: the PHYTOSOME® strategy to improve the bioavailability of phytochemicals. Fitoterapia 81 (5):306–314
57. Andrade CAS, Correia MTS, Coelho LCBB, Nascimento SC, Santos-Magalhães NS (2004) Antitumor activity of Cratylia mollis lectin encapsulated into liposomes. Int J Pharm 278 (2):435–445
58. Priprem A, Watanatorn J, Sutthiparinyanont S, Phachonpai W, Muchimapura S (2008) Anxiety and cognitive effects of quercetin liposomes in rats. Nanomedicine 4(1):70–78
59. Zhou B (2005) Evidence for α-tocopherol regeneration reaction of green tea polyphenols in SDS micelles. Free Radic Biol Med 38(1):78–84
60. Ajazuddin S (2010) Applications of novel drug delivery system for herbal formulations. Fitoterapia 81(7):680–689
61. Ahmad N (2019 Jun) Preparation of a novel curcumin nanoemulsion by ultrasonication and its comparative effects in wound healing and the treatment of inflammation. RSC Adv 9 (35):20192–20206
62. Uner M (2007) Importance of solid lipid nanoparticles (SLN) in various administration routes and future perspectives. Int J Nanomedicine 2(3):289
63. Li H (2009 Feb) Enhancement of gastrointestinal absorption of quercetin by solid lipid nanoparticles. J Control Release 133(3):238–244
64. Harshita MA (2019) Paclitaxel-loaded nanolipidic carriers with improved oral bioavailability and anticancer activity against human liver carcinoma. AAPS PharmSciTech 20(2):87
65. Choubey A (2011) Phytosome-a novel approach for herbal drug delivery. Int J Pharm Sci Res 2(4):807
66. Hegmann T (2007) Nanoparticles in liquid crystals: synthesis, self-assembly, defect formation and potential applications. J Inorg Organomet Polym Mater 17(3):483–508

67. Masson DS (2005) Polyhydroxy alcohols and peach oil addition influence on liquid crystal formation and rheological behavior of O/W emulsions. J Dispers Sci Technol 26(4):463–468
68. dos Santos ODH, Miotto JV, de Morais JM, da Rocha-Filho PA, de Oliveira WP (2005) Attainment of emulsions with liquid crystal from marigold oil using the required HLB method. J Dispers Sci Technol 26(2):243–249
69. Alam MS (2017) Gum ghatti mediated, one pot green synthesis of optimized gold nanoparticles: investigation of process-variables impact using box-Behnken based statistical design. Int J Biol Macromol 104:758–767
70. Barkat MA (2019) Silver nanoparticles and their antimicrobial applications. Curr Nanomedicine 8(3):215–224
71. Barkat Md A (2018) Current progress in synthesis, characterization and applications of silver nanoparticles: precepts and prospects. Recent Pat Antiinfect Drug Discov 13(1):53–69
72. Hasnain MS (2019) Purple heart plant leaves extract-mediated silver nanoparticle synthesis: optimization by box-Behnken design. Mater Sci Eng C Biomimetic Mater Sensors Syst 99:1105–1114
73. Barkat MA (2019) Nanopaclitaxel therapy: an evidence based review on the battle for next-generation formulation challenges. Nanomedicine 14(10):1323–1341
74. Anselmo AC, Mitragotri S (2016) Nanoparticles in the clinic. Bioeng Transl Med 1(1):10–29
75. Ventola CL (2017) Progress in nanomedicine: approved and investigational nanodrugs. P T 42 (12):742–755
76. Luo Y, Cobb RE, Zhao H (2014) Recent advances in natural product discovery. Curr Opin Biotechnol 30:230–237
77. Jadhav P, Shinde N (2014) Herbal nanoparticles: a patent review. Asian J Pharm (AJP) 8(1):58
78. Kuo KW (2006). Water soluble extract from plant of Solanum genus and the preparation process thereof, and pharmaceutical composition containing the water soluble extract. Google Patentss
79. Sahoo SK, Mohanty C (2011). Novel water soluble curcumin loaded nanoparticulate system for cancer therapy. WO Patent 2011101859:A1
80. Singh C. (2004) Novel formulations of digitalis glycosides for treating cell-proliferative and other diseases. Google Patents
81. Kurzrock R, Li L, Mehta K, Aggarawal BB (2011) Liposomal curcumin for treatment of cancer. Google Patents
82. Desai K (2010) Curcumin cyclodextrin combination for preventing or treating various diseases. Google Patents
83. Desai NP, Soon-Shiong P (2010) Nanoparticles of paclitaxel and albumin in combination with bevacizumab against cancer. Google Patents
84. Kato J, Li Y, Xiao K, Lee J, Luo J, Tuscano JM, O'Donnell RT, Lam K (2012) Disulfide cross-linked micelles for the targeted delivery of vincristine to B-cell lymphoma. Mol Pharm 9 (6):1727–1735
85. Einbond L, Redenti S (2013) Growth inhibitory effects of nanoparticles containing triterpene glycosides or triterpenes. Google Patents
86. Frautschy SA, Cole GM (2015) Bioavailable curcuminoid formulations for treating Alzheimer's disease and other age-related disorders. Google Patents
87. Dhawan S, Hooda P, Nanda S (2018) Herbal Nano formulations: patent and regulatory overview. Appl Clin Res Clin Trials Regul Affairs 5(3):159–180
88. Aburdeineh SG, Aburdeineh HG (2012) Methods of lowering blood cholesterol via oral fenugreek seed extract compositions. Google Patents
89. Maurya R, Singh G, Murthy PSN, Mehrotra S, Singh D, Bhargava B, Singh MM (2007) Pharmaceutical composition containing Butea isoflavones for the prevention/treatment of bone disorders and a process for the preparation thereof. WO Patent 2007099432(7)
90. Gupta S (2004) Ion-pair delivery system for cosmetic and pharmaceutical compositions
91. Gupta S (2004) Controlled-release nano-diffusion delivery systems for cosmetic and pharmaceutical compositions. Google Patents

92. Bakthavatchalam R, Blum CA, Brielmann H, Caldwell TM, DeLombaert S, Hodgetts KJ, Zheng X (2008) Carboxylic acid, phosphate or phosphonate substituted quinazolin-4-ylamine analogues as capsaicin receptor modulators. US Patent 7(432):275
93. Matuschek M, Ernst A, Köpsel C, Jager MB, Kleber A, Krohn M, Zinke H (2010). Use of water-dispersible carotenoid nanoparticles as taste modulators, taste modulators containing water-dispersible carotenoid nanoparticles, and, method for taste modulation. Google Patents
94. Spengler E, Dahms G (2006) Composition and method employing membrane structured solid nanoparticles for enhanced delivery of oral care actives. Google Patents
95. Uabundit N (2010) Cognitive enhancement and neuroprotective effects of Bacopa monnieri in Alzheimer's disease model. J Ethnopharmacol 127(1):26–31
96. Dhanasekaran M (2009) Extract selectively decreases amyloid levels in hippocampus of Alzheimer's disease animal model. Phytother Res 23(1):14–19
97. Henrotin Y (2010) Biological actions of curcumin on articular chondrocytes. Osteoarthr Cartil 18(2):141–149
98. van Beek TA (2002) Chemical analysis of Ginkgo biloba leaves and extracts. J Chromatogr A 967(1):21–55
99. Lalotra S, Vaghela JS (2019) Scientific reports of medicinal plants used for the prevention and treatment of neurodegenerative diseases. Pharm Biosci J 7(1):15–25
100. Divino da Rocha M (2011) The role of natural products in the discovery of new drug candidates for the treatment of neurodegenerative disorders II: Alzheimer's disease. CNS Neurol Disord Drug Targets 10(2):251–270
101. Vestergaard M (2010) Cholesterol, lipids, amyloid beta, and Alzheimers. Curr Alzheimer Res 7(3):262–270
102. Imbimbo B (2008) Therapeutic potential of γ -secretase inhibitors and modulators. Curr Top Med Chem 8(1):54–61
103. Rojas P (2011) S-Allylcysteine, a garlic compound, protects against oxidative stress in 1-methyl-4-phenylpyridinium-induced Parkinsonism in mice. J Nutr Biochem 22 (10):937–944
104. Deleu D (2002) Clinical pharmacokinetic and pharmacodynamic properties of drugs used in the treatment of Parkinson's disease. Clin Pharmacokinet 41(4):261–309
105. Aguiar LMV (2006) Neuroprotective effects of caffeine in the model of 6-hydroxydopamine lesion in rats. Pharmacol Biochem Behav 84(3):415–419
106. Suganuma H (2002) Effect of tomato intake on striatal monoamine level in a mouse model of experimental Parkinson's disease. J Nutr Sci Vitaminol 48(3):251–254
107. Radad K (2009) Thymoquinone protects dopaminergic neurons against MPP and rotenone. Phytother Res 23(5):696–700
108. Lahaie-Collins V (2008) Sesamin modulates tyrosine hydroxylase, superoxide dismutase, catalase, inducible no synthase and interleukin-6 expression in dopaminergic cells under Mpp -induced oxidative stress. Oxidative Med Cell Longev 1(1):54–62
109. RajaSankar S (2009) Ashwagandha leaf extract: a potential agent in treating oxidative damage and physiological abnormalities seen in a mouse model of Parkinson's disease. Neurosci Lett 454(1):11–15
110. Kyung KS (2006) 6-Shogaol, a natural product, reduces cell death and restores motor function in rat spinal cord injury. Eur J Neurosci 24(4):1042–1052
111. Jose S, Sowmya S, Cinu TA, Aleykutty NA, Thomas S, Souto EB (2014) Surface modified PLGA nanoparticles for brain targeting of Bacoside-a. Eur J Pharm Sci 63:29–35
112. Raval N, Barai P, Acharya N, Acharya S (2018) Fabrication of peptide-linked albumin nanoconstructs for receptor-mediated delivery of asiatic acid to the brain as a preventive measure in cognitive impairment: optimization, in-vitro and in-vivo evaluation. Artifl Cells Nanomedicine Biotechnol 46(sup3):S832–S846
113. Cheng KK, Yeung CF, Ho SW, Chow SF, Chow AHL, Baum L (2013) Highly stabilized Curcumin nanoparticles tested in an in vitro blood–brain barrier model and in Alzheimer's disease Tg2576 mice. AAPS J 15(2):324–336

114. Wang P, Cao X, Chu Y, Wang P (2019) Ginkgolides-loaded soybean phospholipid-stabilized nanosuspension with improved storage stability and in vivo bioavailability. Colloids Surf B Biointerfaces 181:910–917
115. Sathya S, Shanmuganathan B, Manirathinam G, Ruckmani K, Devi KP (2018) α-Bisabolol loaded solid lipid nanoparticles attenuates Aβ aggregation and protects Neuro-2a cells from Aβ induced neurotoxicity. J Mol Liq 264:431–441
116. Pan L, Zhou J, Ju F, Zhu H (2018) Intranasal delivery of α-asarone to the brain with lactoferrin-modified mPEG-PLA nanoparticles prepared by premix membrane emulsification. Drug Deliv and Transl Res 8(1):83–96
117. Freag MS, Saleh WM, Abdallah OY (2018) Self-assembled phospholipid-based phytosomal nanocarriers as promising platforms for improving oral bioavailability of the anticancer celastrol. Int J Pharm 535(1–2):18–26
118. Yu H, Teng L, Meng Q, Li Y, Sun X, Lu J, Lee R, Teng L (2013) Development of liposomal Ginsenoside Rg3: formulation optimization and evaluation of its anticancer effects. Int J Pharm 450(1–2):250–258

Nanophytomedicine Ethical Issues, Regulatory Aspects, and Challenges

Roohi Mohi-ud-din, Reyaz Hassan Mir, Faheem Hyder Pottoo, Gifty Sawhney, Mubashir Hussain Masoodi, and Zulfiqar Ali Bhat

Abstract

Herbal medicines have been used around the globe since times immemorial. Nanophytomedicine is an emerging domain of medicine which helps to encapsulate the active phytoconstituents from medicinal plants and treat various diseases such as cancer and diabetes. So it is very crucial to identify the potential active phytoconstituents and encapsulate them to treat diseases without any major side effects. Nanophytomedicine has led to increment in the bioavailability of water and therapeutic compounds which are insoluble by utilizing drug formulations of various combinatorial. Increasing evidence has suggested that phyto-bioactive compounds show enhanced benefits in curing various diseases. Curcumin, resveratrol, ginsenosides, quercetin, and catechin are phyto-derived bioactive compounds with important roles in the prevention and treatment of neurological diseases. However, in vivo studies suggest that their concentrations are very low to cross blood–brain barrier thereby it limits bioavailability, stability, and dissolution at target sites in the brain. To overcome these problems, nanophytomedicine with the controlled size of 1–100 nm is used to maximize

R. Mohi-ud-din · Z. A. Bhat (✉)
Pharmacognosy and Phytochemistry Lab, Department of Pharmaceutical Sciences, School of Applied Sciences and Technology, University of Kashmir, Srinagar, Jammu and Kashmir, India
e-mail: roohidin@kashmiruniversity.net; zabhat@kashmiruniversity.ac.in

R. H. Mir · M. H. Masoodi
Pharmaceutical Chemistry Division, Department of Pharmaceutical Sciences, School of Applied Sciences and Technology, University of Kashmir, Srinagar, Jammu and Kashmir, India

F. H. Pottoo
Department of Pharmacology, College of Clinical Pharmacy, Imam Abdulrahman Bin Faisal University, Dammam, Saudi Arabia

G. Sawhney
Inflammation Pharmacology Division, CSIR-Indian Institute of Integrative Medicine, Jammu, Jammu and Kashmir, India

S. Beg et al. (eds.), *Nanophytomedicine*,
https://doi.org/10.1007/978-981-15-4909-0_10

efficiency. Nanosizing of phyto-bioactive compounds enhances the permeability into the brain with maximized efficiency and stability. Several nano-delivery techniques, including solid lipid nanoparticles, nanostructured lipid carriers, nanoliposomes, and nanoniosomes can be used for controlled delivery of nano-bioactive compounds to brain. Nanocompounds, such as ginsenosides (19.9 nm) synthesized using a nanoemulsion technique, showed enhanced bioavailability in the rat brain.

Keywords

Nanophytomedicine · Ethics · Regulatory affairs · Herbal medicines · Nanomedicine · Nanotechnology

1 Introduction

The therapeutic benefits of herbals, an ancient form of medicine, were first recognized by Greek pharmacist Galen for treating a variety of ailments (129–200 AD) [1]. For the development of drugs, natural products played a pivotal role in the discovery of lead molecules for treating various diseases [2]. Indeed 60% of the currently available drugs in the market are directly or indirectly procured from natural products [3, 4]. Moreover, 80% population in African and Asian countries are dependent on phyto-drugs in primary health care [5]. As a result of good efficacy and fewer adverse effects of phytomedicines, their usage has increased manifolds as compared to allopathic medicines (Fig. 1). However, less bioavailability and poor

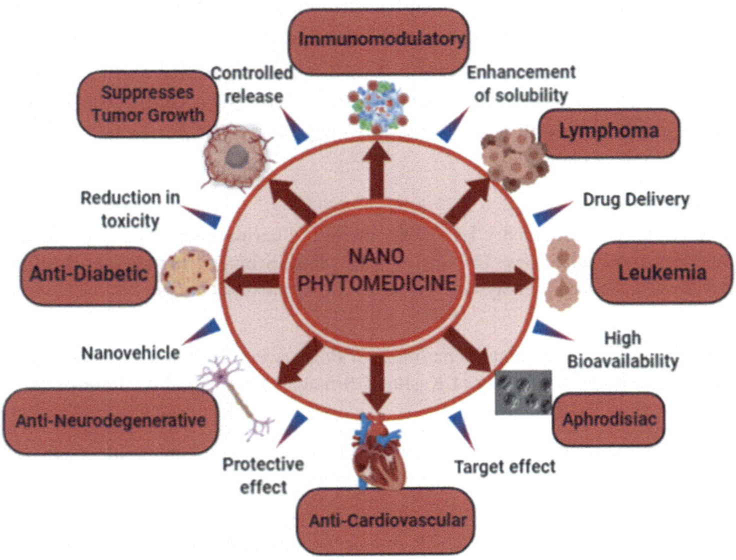

Fig. 1 Nanophytomedicines for treating various disorders

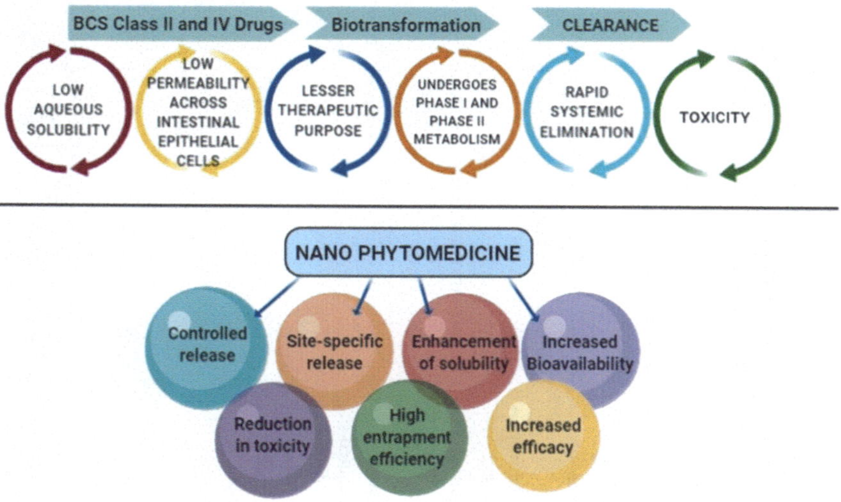

Fig. 2 Nanophytomedicine enhances the bioavailability and bioactivity of phytomedicine

solubility in water are the major factors (Fig. 2) that limit their use in the treatment and prevention of various disease [5–8]. So, material nanoscience in the form of nanophytomedicine overcomes the problems including toxicity, solubility, degradation, absorption, and biological response of phytomedicine [9, 10]. Nanophytomedicines are generally prepared from standard extracts or active phytoconstituents [11, 12]. The biological activity and therapeutic activity of several phytomedicines have been improved to manifolds by integrating nanotechnology approaches and transforming them into nanophytomedicine [13–15]. In Table 1, a detailed list of examples has been added indicating recent developments in nanophytomedicine.

The field of nanotechnology in recent years has been experiencing dynamic development. In each country, research and development expenditure in this sector is increasing. Nanotech applications are used in fields such as medicine, telecommunications, electricity, food industry, information technology, astronautics, and communication. Nanotechnology is called a technology of dual use because it offers both the opportunities for advancement and human development, but also a serious threat to human health and life and the environment [16]. It is an extremely young field and it is not always known the impact of nanomaterials on human bodies and the environment. Nanomaterials can not only increased biological activity loaded drugs but also increased response from loaded high molecular weight biological macromolecules such as proteins, DNA miRAs because they infiltrate biological membranes quickly [17]. Therefore, nanoparticles may have toxic properties that threaten humans and animals. Apparently, there is no scientific way to track the existence and effect of nanoparticles and nanomaterials, for example, in the atmosphere, because of the lack of supervision in nanotechnology. While the European Union (EU) has adopted nanotechnology regulations since 2004, there is

Table 1 Recent developments in nanophytomedicine

Medicinal plant	Active ingredient	Nano-formulation	Method/nanotechnology approaches	Rationale	Targets studied	References
T. brevifolia	Paclitaxel	EndoTAGR -I (SB05)	Cationic liposomal paclitaxel		Breast cancer/pancreatic cancer, and head and neck squamous cell carcinoma	[28–31]
		LEP-ETU	Liposomal paclitaxel, NeoLipid® technology	Potential to regulate multidrug resistance	Ovarian/breast/lung cancer	
Curcuma longa	Curcumin	Glyceryl Mono-oleate nanoparticles	High-pressure homogenization method	Enhance bioavailability in brain tissue	Glioblastoma therapy	[32]
Tripterygium wilfordii Hook F	Celastrol	Celastrol nanoparticles	PEG-block-poly (€-caprolactone) nano polymeric micelles	To improve hydrophilicity	Retinoblastoma therapy	[33]
Ginkgo biloba	Bilobalide and Ginkgolides A, B, C	PEGylated nanoparticles	Co-encapsulation	Synchronized and sustained release of components		[34]
Scutellaria baicalensis Georgi	Baicalein	Nanocrystals	Anti-solvent recrystallization followed by high-pressure homogenization	To enhance bioavailability		[32]
Boswellia caraterii and *Commiphora myrrha*	Myrrh and essential oils	Solid nanoparticles	High-pressure homogenization	To improve hydrophilicity	Antitumor efficacy	[35]

Salvia miltiorrhiza Bunge	Cryptotanshinone	Solid lipid nanoparticles	Ultrasonic and high-pressure homogenization	To enhance bioavailability		[36]
Kaempferia parviflora	Extract of Kaempferia parviflora	Solid lipid nanoparticles	Oil in water microemulsion	To enhance transdermal permeability		[37]
Strychnos nux-vomica	Brucine	Polyacetic acid nanoparticles	Solvent diffusion method	Sustained release		[38]
Panax notoginseng	Panax notoginsenoside	Panax notoginsenoside-loaded core-shell hybrid liposomal vesicles	Solvent evaporation method	To enhance its protective effects	Protective effects on cerebral ischemia	[39]
Artemisia annua L.	Artemisinin	Nanocapsules	self-assembly procedure	To enhance bioavailability	Anticancer therapy	[40]

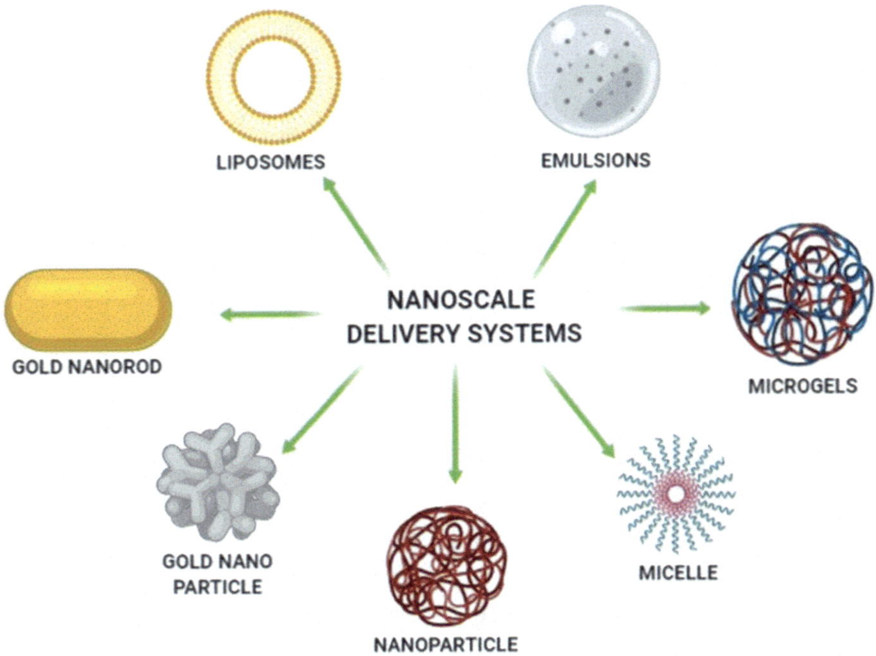

Fig. 3 Various types of nano-delivery systems

still no appropriate regulation. The EU legislation aims at providing the public with exposure to creative nanotechnology technologies, privacy, health and environmental protection. The topic of the ethics in the area of nanosciences and nanotechnologies is also constantly being discussed. Relatively complexity involved and crossed-talk mechanism of various disorders, synthetic drugs proven to exhibit limited clinical efficacy motivated rationalizing various plants which are although undocumented in scientific texts but bearing long history of clinical evidences in different populations throughout human development to these alleviate diseases [18–21]. Plants have been known and used as herbal medicinal products in different populations throughout human development, beginning when people were learning to choose plants for food and alleviate diseases [14, 15, 22, 23]. Such scientific advances have revolutionized the availability of drugs. The new systems for drug delivery not only improve the potency of active components but also enable certain components to be reintroduced that have been scrapped because they were not successful in formulation [11, 12] (Fig. 3).

Furthermore, this method is even more appealing because of its ability to improve novel drugs, such as increasing the selectivity and potency, shielding them from thermal or photo-degrading, minimizing side effects, and testing the development of active compounds prior to being put on the market or used therapeutically [24–26]. Increasing interest in nanotechnological progress has become increasingly apparent in pharmaceutical industries, because such developments bring benefits, such as modified release schemes, and the potential for developing previously impossible new formulations (by several aspects related to the active constituents) [27]. Although the advances of nanotechnology are useful for several drugs, some of the drawbacks must be highlighted. Clinical experts have discussed other negative factors, such as high costs and scaling problems and simple nanoparticles on inhalation can lead to dangerous lung disorders and often other diseases contributing to shifts in homeostasis or even death. Table 2 summarizes bioavailability of plant-based nano-bioactive compounds and their production methods. Also some of the biological effects of exemplar homeopathically manufactured (nano)phytomedicines are represented in Table 3.

2 Ethical Issues and Regulatory Aspects Associated with Nanophytomedicine

As a rule, for the development of any product, the clinical research involving nanophytomedicines has to be placed under similar ethical standards in accordance with the national and international ethical regulations and guidelines. The plan and method of research must follow ethical guidelines. Proper discussion, identification, and addressing of all the issues pertaining to ethics in a manner acceptable under ethical regulations and guidelines are necessary especially those issues involving cognition and innate characteristics of nanophytomedicine. There are various ethical issues to be addressed regardless of the advancement in research in nanotechnology and its heavy promotion. To some extent, these ethical issues are comparable with those confronted by most of the developing technologies. Nevertheless, with the development of phytomedicines, a large number of these ethical issues are being discussed before the actual beginning of this technology. The result of this discussion was that it has a significant influence on the day to day life of a person. Ethics is led by the science at some occasions and as a result, it creates a condition in which ethical perplexity arises as ethical and regulatory views are significant in practicing medicine, the same is applicable to nanophytomedicine. The FDA is devising explicit regulations pertinent to nano-biotechnology products. That said, the FDA will regulate the development of pharmaceuticals which comprise of nanoparticles and techniques of drug delivery as it does for any other biopharmaceutical product. It is very difficult to take care of all the ethical problems pertaining to new technology like nanophytomedicine. However, it would be advantageous if many of these are addressed beforehand. The ethical issues that are usually encountered by nanophytomedicines include the particle size, its influence on human health, environment safety, potential unexpected after come and harmful influenced by the use of nanophytomedicine. To conduct research on social, legal, and ethical issues emerging from the introduction of nanotechnology, the universities have been provided with US Federal grants.

Table 2 Bioavailability of plant-based Nano-bioactive compounds and their production methods

Bioactive compounds	Production methods	Particle size (nm)	Experiment models	Bioavailability	References
Nanocurcumin	Alginate-curcumin nanocomposite	11.3	Drosophila model	Increase bioavailability and decrease oxidative stress and apoptosis in Parkinson's disease	[41]
	Solid lipid nanoparticle	190	Balb/c mice	Increase concentrations in lungs	[42]
	PLGA nanoparticles	100–200	HeLa cells	Increase anticancer efficiency	[43]
	PLGA nanoparticles	80–120	Human neuroblastoma SK-N-SH cells	Decrease neurons against oxidative damage in Alzheimer's disease	[44]
	PLGA nanoparticles	158	Sprague Dawley rats	Increase intravenous bioavailability	[45]
	PLGA nanoparticles	NA	Sprague Dawley rats	Increase neuroprotective effect	[46]
Nanoginsenosides	Nanoliposome	150	L929 cells	Enhance the survival rate of H_2O_2-damaged cells	[47]
	Ginsenoside compound K-bearing glycol chitosan conjugate nanoparticles	255 and 296	Cancer cell lines	Exhibited higher cytotoxicity to HT29, HepG2, and HT22 cancer cells	[48]
Nanoresveratrol	Nanocapsule	241	Male Wistar rats	Enhanced bioavailability in Alzheimer's disease	[49]
	PCL-PeG polymeric micelles	100	PC12 cells	Enhanced bioavailability in Alzheimer's disease	[50]
	Eudragit e100	73.8	Male Wistar rats	Decrease oxidative stress and prevent chronic liver disease	[51]
	PeG-PLA-resveratrol conjugates	150	Rat C6 and Human U87 glioma cells	Increase antitumor activity	[52]
	Resveratrol-loaded PeG-PLA polymeric NPs	120–233	Cultured CT26 colon cancer cells *in vitro* and in CT26 tumor-bearing mice in vivo	Higher antitumor activity	[53]

Nanocatechin	Nanoencapsulation	432–440	Swiss outbred mice	Increase bioavailability	[54]
	Tea polyphenol-loaded chitosan nanoparticles	400–452	HepG2 cells	Increase antitumor	[55]
	Nanoliposome	71.7	In vitro	Increase bioavailability	[56]
	Epigallocatechin-3-gallate gold nanoparticle	64.7	B16F10 murine melanoma cells	Improved anticancer efficacy	[57]
	Epigallocatechin-3-gallate- loaded nanoparticles prepared from chitosan and polyaspartic acid	102	Oral administration to rabbits	Decrease atherosclerosis	[58]
Nanoquercetin	Solid lipid nanoparticles	200	Male Wistar rats	Increase brain antioxidant capacity	[59]
	Solid lipid nanoparticles	155.3	Male Wistar rats	Increase sustained release	[60]
	Nanosuspension	430	In vitro	Higher antioxidant activity	[61]
	Nanoliposomes	62.3–191.5	C6 glioma cells	Anticancer activity	[62]
Nanolycopene	Nanostructured lipid carrier	150–160	In vitro	Higher antioxidant activity	[63]
	Nanoemulsions	100–200	In vitro	Higher bioavailability	[64]

Abbreviations: *NA* not available, *NP* nanoparticle, *PCL* poly-caprolactone, *PeG* polyethylene glycol, *PLGA* polylactic-co-glycolic acid

Table 3 Biological effects of exemplar homeopathically-manufactured nanophytomedicines

Homeopathic phytomedicine	Biological effects	References
Aconitum napellus	In combination potentized medicine Canova, induced macrophage biochemical activity resulting in reactive oxygen species and nitric oxide production	[65]
Aloe vera	3X, 6X, 12X, 30X potencies as part of a combination homeopathic product versus 20% ethanol control stimulated release of pro- and anti-inflammatory cytokines by human white blood cells	[66]
Arnica montana	Combination homeopathic product including Arnica 3× potency blocked experimentally-induced edema in rat paws	[67]
Asclepias vincetoxicum	6×, 10×, 30× potencies as part of a combination homeopathic product demonstrated antiviral activity independent of activation of the cellular interferon system	[68]
Bryonia alba	In combination potentized medicine Canova, induced macrophage biochemical activity resulting in reactive oxygen species and nitric oxide production	[65]
Calendula officinalis	Combination homeopathic product including Calendula 4× potency stimulated fibroblast migration and promoted experimental wound closure significantly better than succussed control solvent. Unsuccussed solvent had no effect on cell migration	[69]
Gelsemium sempervirens	Reduced anxiety-related behaviors in open field testing at 5C, 9C, and 30C potencies comparable to conventional anxiolytic drug buspirone in mice	[70]
Hydrastis canadensis	200C potency significantly increased lifespan of mice with transplanted Ehrlich Ascites Carcinoma and Dalton's Lymphoma Ascites induced-tumors by 69.4% and 1 M (1000C) potency reduced metastases of melanoma-affected mice	[71]
Hypericum perforatum	Combination homeopathic product including Hypericum 4× potency stimulated fibroblast migration and promoted experimental wound closure significantly better than succussed control solvent. Unsuccussed solvent had no effect on cell migration	[69]
Lycopodium clavatum	200C potency reduces cytogenetic damage from exposure to chemical carcinogens in mice	[72]
Phytolacca decandra	Homeopathic mother tincture reduces proliferation, activates caspase-mediated signaling and reactive oxygen species generation with DNA damage in A375 cancer cells but not healthy lymphocytes	[73]
Pulsatilla nigricans	3× potency produced greater anxiolytic effects than 6× potency or bulk diazepam drug in mice on a common measure of anxiety (elevated plus maze and open field test)	[74]

2.1 Social Impact of Nanophytomedicines

The development of any new scientific technology has been mostly linked with ethical perplexities, especially with respect to regulation and containment. More ethical dilemmas are likely to arise from the technologies having a significant influence on society. Nanophytomedicines have the ability to influence society because of its interdisciplinary nature and capability in treating many diseases. On the one hand, active constituents in phytomedicine being of small size possess unmatched attributes but unluckily this same advantage may lead to some difficulties. It becomes very difficult to remove the particles by current filters because of their small size and thus, causing deliberate dissemination of the unrealized ability of nanophytomedicine. Moreover, these minute particles or structures cannot be seen by the naked eye. This raises an important query for how nanophytomedicine as a technology be contained as such. The impact of nanophytomedicine on the environment and human health is an important issue with respect to health, safety, and ethics [75]. How will the nanophytomedicine interact with the environment or the human body? Will these have toxic effects or cause unanticipated reactions specifically in contrast to their macroscopic analogs? These are the questions arising pertinent to nanophytomedicine that are still unanswered and as a consequence, people remain careful particularly after the incidents like Mile Island and Love Canal depicting the harm caused to the environment. During the process of production of nanophytomedicine, workers are at risk of getting exposure to higher doses of nanophytomedicine by means of ingestion, inhalation or penetration through the skin. The toxicity of nanophytomedicine particles being unable to predict can be more than that of their macroscopic analogs. Exposure at the workplace should be tracked and regulations governing exposure levels for manufactured nanophytomedicines should be reviewed at regular intervals.

2.2 Ethics in Nanophytomedicine

All new technologies are governed by some ethical regulations and nanobiotechnology cannot be exempted from the same. Even though nanotechnology has not brought up any new ethical issues, it becomes important to take care of these considerations during the development and application of nano-bio technologies to medicine. Lately, with an eye to lessen the adverse effects of nanotechnology on the environment and the human health and to avoid aggravated response from people, ethical, social, and legal queries arising due to the incorporation of nanotechnology in medicine have been reevaluate [76, 77].

A need has arisen for the development of largely harmonized and well-instructed appraisal [78]. Nanomedicine in general encounters various problems which include:

1. The ethical limitation to providence apprehended as reasonable risk assessment;
2. The diminution of ethics to cost/benefit analysis; and

3. The deficiency of obviousness in technique with technology and of human nature with the human condition [79]. Once these things are explicated, it would be evitable to address some rational and immaterial queries regarding nano-bio technologies.

In 2006, the Nanoethics Group, situated in Cal PolyState University (San Luis Obispo, CA), was furnished grants through the National Science Foundation to study ethical issues concerning human advancement and nanotechnology. The role of nanotechnology in the advancement practices is not clearly evident. Advancement means the use of hormones or esthetic surgeries or distinct procedures to improve the human performance (mental or physical) or appearance. The group is solicitous about the expansion of this new technology that it may lead to certain unusual occurrences such as sophisticated cybernetic organs and computers introduced inside the brains, which may produce ethical concerns. During the year 2007, the Nanoethics Group put forth series of papers to tackle both pressing and far-off issues regarding the influence of nanotechnology on people [80]. It deals with a complete series of issues confronting nanotechnology, especially with respect to advantages, possible negative outcomes, environment, state of well-being of humans, human improvement, seclusion, defense, democratism, imparting knowledge, humanitarianism, molecular production, space traveling, artificial intelligence, increasing life span, etc. [80].

A draft report was delivered by EGE, the European Group on Ethics in Science and New Technologies in 2007 which realizes the ability of nanomedicine to promote new diagnostic techniques and therapies. The Nanoethics group proposed that mechanisms should be developed to authenticate the safety of products and equipment used in nanomedicine, and convenes the relevant people with authority to conduct an appropriate assessment of the risks and safety of nanomedicine. It also suggests that an EU website should be created covering ethics and nanomedicine, where countrymen locate information and place their queries to researchers. At present, the process of considering ethical issues implicated in nanomedicine are associated with risk evaluation broadly, somatic-cell versus germline-cell therapy, the improvement of human potential, managing negative outcomes of engineered nanomaterials, research on embryonic stem cells of human and the degree to which nanoparticles and nanomedicine may be toxic, uncontrolled action and ability of nanoparticles gathering together by themselves. For example, the realization of the toxic nature of nanoparticles in relation to mammalian germline stems cells has led to a large number of queries about the biosafety of nanomaterials. During their research, a cell line possessing the properties of spermatogonial stem cell was employed to examine the in vitro toxic nature of different categories of nanoparticles.

2.3 Regulatory Aspects of Nanotechnology-Based Phytomedicines

Nano-biotechnology is liable to bring up legal issues as any different recent technology. Yet no law has been formulated with respect to nanotechnology. Due to safety issues pertaining to exposure of nanoparticles into the surrounding environment and the workplace, and in addition, implementation in human health care, there is a fear of disagreement among the masses in the time ahead. Hence, it becomes important to follow existent laws as successful examples regarding the usage of drugs and therapies. Even so, this domain of technology needs explicit edicts for the reason that a lot of difficulties will emerge in the time ahead that have not been regulated up to this time. Moreover, human safety has to be assured as the main goal and this can be accomplished by means of a considerable legal ruling. Some fear that the new EU's chemical policy, REACH, i.e., registration, evaluation, authorization, and restriction of chemicals might be adopted as reference for the regulation of nanotechnology, which might imply a qualified shift of the burden of proof for safety, from the authorities to the manufacturer. An overview of current and future EU regulation of nanotechnology, with some comparisons between the EU and US regulatory frameworks, has been published. The scope of nanotechnology-based phytomedicines for human use reflects current thinking and initiatives taken by the European Medicines Agency (EMEA) following recent development of nanotechnology-phytomedicines. Nanotechnology is an emerging scientific research field with wide applicability and, in the context of medical science, is expected to contribute in developing a more proactive paradigm for the diagnosis and therapy of disease. Medicinal products containing nanoparticles have already been authorized both in EU and the US under existing regulatory frameworks. Despite of regulating wide range of products by FDA including foods, cosmetics, drugs, devices, and veterinary products, some of which may utilize nanotechnology or contain nanomaterials, the FDA yet has not established its own formal definition. The FDA approval is essential for clinical applications of new technologies and substantial regulatory problems may be encountered in the approval of nanotechnology-based products. The FDA also has an internal Nanotechnology Task Force, which deals with determining regulatory approaches that encourage the continued development of innovative, safe, and effective FDA-regulated products that use nanotechnology materials [76–80].

3 Challenges Encountered in Nanophytomedicine Product Development

Despite the deficiency of regulatory standards for testing of nanoparticle-based medicines, exertions are being made to generate an unmatched collection of therapeutic agents. Nanoscience and technology are yet dealing with various new confrontations regardless of significant advancement in nanotechnology and increase in various commercialized products containing nanoparticles. The concept of nanophytomedicine has gained considerable attention in the last few decades and is the current thrust area of dug administration progressing to and fro with a wider objective of targeted drug delivery including minimization of side effects [81, 82]. The Herbal drugs or Phytomedicines are involved in the conventional medicinal system from a very long time. Globally, Herbal medicine is accepted as an optional system to the present conventional system of medication. They are not only utilized as a therapeutic agent but also as nutraceuticals and functional foods too. Novel drug delivery system has opened new doors during the development of phytomedicine for better therapeutic effects and in minimizing their toxicity [82–84].

The major objective concerned with the development of nanophytomedicine remained the target-specific delivery of phytomedicine to achieve maximum therapeutic benefits and to minimize the risk of drug toxicity. Nanotechnology has proven them as a versatile drug carrier for phytomedicine. Certain important confrontations have been encountered during the development and employment of nano-drug delivery systems in phytomedicine based nano-formulation used for the prevention and treatment of various diseases [85–89]. Following hindrances in growth can be the barriers for growth, as enumerated below, if overcomed can lead to significant developments in the area of health care and medicine.

a. First and foremost production of active constituents of phytomedicine via process of screening, isolation, and identification is in itself a major challenge as these phytoconstituents are found in very low amounts in the plants. Moreover, their yield tends to vary depending on various external and internal factors, thus took months to get a desired quantity of active ingredient for preclinical development [90–92].
b. To prepare the nanoparticles for targeted delivery of phytomedicines to their specific targets without causing any harm to rest of the body. Various studies have been conducted in vitro for testing phytomedicines against diseases. Hence, it is needed to conduct more research over in vivo animal models in order to explicate the effectiveness of phytomedicines.
c. To increase the production of nanomaterial employed for the formation of nanoparticle. The workability of these procedures leads to introduction of new therapeutic techniques into the market very soon. The main confrontation during the increment of formulation is to limit the size of the nanoparticle. Consequently, there is variation between the batches. To overcome these problems, biosurfactant molecules prove to be useful in the large-scale production as well as technology growth.
d. Certain confrontations are being faced while converting these technologies into therapies. This arises due to interactions between nanoparticles and living organisms and should be comprehended and brought under control. Proper investigations yet need to be conducted to study the association of the structure and functions of the nanoparticles. It has been observed several times that these nanoparticles adsorb the protein part of plasma causing impedance in the defense

mechanisms of the body. Genotoxicity may also be caused when these nanoparticles lead to the formation of free radicals.

e. Another crucial confrontation is the characterization of these nanomaterials employing several procedures such as fluorescence based assays, cytotoxicity studies, etc. which are hampered because of the physical and chemical characteristics of nanomaterials. In the time ahead, more attention needs to be paid on drug loading efficiency and lowering the disposal of doses. Further investigations need to be conducted for the immunotoxicity of nanoformulations arising due to unanticipated immune response after the administration [93, 94].
f. Assessment of toxicological profile and safety of nanophytomedicine. Characterization of synthesized nanoparticles by means of its toxicity and safety profile is getting challenging. Due to reduced particle size and increased surface area nanomedicines have gained appealing potential to reduce the dose frequency and bioavailability enhancement. But, the main concern still remains there regarding effectiveness and safety as there is distress as about how much that nanoformulation should be taken. If any formulation makes pharmacokinetic profile better, but it could not deal with the safety and produces toxicity within the body can be a limitation of nanomedicine [93–95].
g. Although, the role of nanophytomedicine for enhanced drug delivery is a major issue the possible toxic effects of these nanoparticles cannot be tolerated. There are various types of nanoparticle like gold nanoparticles, TiO2, CdCl2, etc. where the cytotoxicity can be provoked at any stage. Certain examples like carbon nanotube induced pulmonary toxicity put a question mark on the safety profile of these nanoparticles for intended human use. Various nanoparticles comprising formulation intended for skin application have caused significant dermal inflammation, pigmentation, and other associated complications [96, 97].

   As nanomedicines have got complex nature leading to multiform toxicity, it is not easy to propose an outline pertaining to special safety guideline for a particular nanoparticle. So, it becomes compulsory to provide a safety protocol, experimental proof, and comprehensive preclinical testing. At present, there is a crucial need for developing nanotechnology together with the phytomedicine has safety guideline by the government to keep strict check the side effects of these nanophytomedicines [96, 98].
h. Focus should be moved in current research towards the possibility of designing and development of nanophytomedicine systems with multiple-use in order to achieve various biological and therapeutic requirements and in vivo studies of their formulations.
i. Due to use of various techniques and equipment in nanomedicine characterization, their cost is much higher than the conventional ones. Besides, there is upgraded level of safety as compared to traditional, but upgradation in efficacy levels is not usually recommended by health care professionals [89, 99].

4 Conclusion

Nanoparticle-based therapeutics has got capacity to overcome biological barriers and to deliver hydrophobic drugs and biologics effectively to the target sites of disease. But there is need of scrupulous design development, engineering techniques, reproducible scale-up, and manufacturing process to accomplish a consistent and uniform product due to the complexity of nanoparticles as multi-component 3D structures. Thus, even a minor change in multiple parameters can influence the safety profile and efficacy of nanoparticle-based medicines, so there is need of careful examination in preclinical and clinical studies. Finally, nanoparticle-based medicines may have to represent additional development challenges and regulatory considerations compared with conventional medicines.

Also, because of the limited data on the physicochemical properties of nanoscale materials in biological systems, clinical trials of nanomedicine products present some unique challenges related to risk minimization, management, and communication involving human subjects. Nanotechnology-based products although are capable of overcoming the limitations of traditional methods. But, the major challenges are yet to prevail over its toxicity, environmental hazards, production cost, and accessibility to the un-reachable at far-off areas.

References

1. Cheng Z, Zhen C (2004) The Cheng Zhi-Fan collectanea of medical history. Peking University Medical Press, Beijing
2. Newman DJ, Cragg GM (2007) Natural products as sources of new drugs over the last 25 years. J Nat Prod 70(3):461–477
3. Anagnostaki EE, Zografos AL (2012) "Common synthetic scaffolds" in the synthesis of structurally diverse natural products. Chem Soc Rev 41(17):5613–5625
4. Newman DJ (2008) Natural products as leads to potential drugs: an old process or the new hope for drug discovery? J Med Chem 51(9):2589–2599
5. Khare CP (2004) Indian herbal remedies: rational Western therapy, ayurvedic, and other traditional usage. Springer, New York
6. Rasoanaivo P et al (2011) Whole plant extracts versus single compounds for the treatment of malaria: synergy and positive interactions. Malar J 10(1):S4
7. Javed MN, Alam MS, Waziri A, Pottoo FH, Yadav AK, Hasnain MS, Almalki FA (2019) QbD applications for the development of nanopharmaceutical products. In: Pharmaceutical quality by design. Academic Press, New York, pp 229–253
8. Javed MN, Kohli K, Amin S (2018) Risk assessment integrated QbD approach for development of optimized bicontinuous mucoadhesive limicubes for oral delivery of rosuvastatin. AAPS PharmSciTech 19(3):1377–1391
9. Ghosh V et al (2013) Antibacterial microemulsion prevents sepsis and triggers healing of wound in wistar rats. Colloids Surf B: Biointerfaces 105:152–157
10. Rajendran R et al (2013) Development of antimicrobial cotton fabrics using herb loaded nanoparticles. Carbohydr Polym 91(2):613–617
11. Alam MS, Javed MN, Pottoo FH, Waziri A, Almalki FA, Hasnain MS, Garg A, Saifullah MK (2019) QbD approached comparison of reaction mechanism in microwave synthesized gold nanoparticles and their superior catalytic role against hazardous nitro-dye. Appl Organomet Chem 33(9):e5071

12. Alam MS, Garg A, Pottoo FH, Saifullah MK, Tareq AI, Manzoor O, Mohsin M, Javed MN (2017) Gum ghatti mediated, one pot green synthesis of optimized gold nanoparticles: Investigation of process-variables impact using Box-Behnken based statistical design. Int J Biol Macromol 104:758–767
13. Hasnain MS, Javed MN, Alam MS, Rishishwar P, Rishishwar S, Ali S, Nayak AK, Beg S (2019) Purple heart plant leaves extract-mediated silver nanoparticle synthesis: optimization by Box-Behnken design. Mater Sci Eng C 99:1105–1114
14. Sharma S, Javed MN, Pottoo FH, Rabbani SA, Barkat M, Sarafroz M, Amir M (2019) Bioresponse inspired nanomaterials for targeted drug and gene delivery. Pharma Nanotechnol 7(3):220–233
15. Mishra S, Sharma S, Javed MN, Pottoo FH, Barkat MA, Alam MS, Amir M, Sarafroz M (2019) Bioinspired nanocomposites: applications in disease diagnosis and treatment. Pharma Nanotechnol 7(3):206–219
16. Dorocki S, Kula A (2015) Przestrzenne zróżnicowanie rozwoju nanotechnologii w Europie. Prace Komisji Geografii Przemysłu Polskiego Towarzystwa Geograficznego 1(29):27–41
17. Pottoo FH, Barkat MA, Ansari MA, Javed MN, Jamal QM, Kamal MA (2019) Nanotechnologoical based miRNA intervention in the therapeutic management of neuroblastoma. Sem Cancer Biol. https://doi.org/10.1016/j.semcancer.2019a.09.017
18. Pottoo FH, Tabassum N, Javed MN, Nigar S, Sharma S, Barkat MA, Alam MS, Ansari MA, Barreto GE, Ashraf GM (2020) Raloxifene potentiates the effect of fluoxetine against maximal electroshock induced seizures in mice. Eur J Pharm Sci. https://doi.org/10.1016/j.ejps.2020.105261
19. Nigar S, Pottoo FH, Tabassum N, Verma SK, Javed MN (2016) Molecular insights into the role of inflammation and oxidative stress in epilepsy. J Adv Med Pharma Sci 10(1):1–9
20. Pottoo FH, Javed M, Barkat M, Alam M, Nowshehri JA, Alshayban DM, Ansari MA (2019) Estrogen and serotonin: complexity of interactions and implications for epileptic seizures and epileptogenesis. Curr Neuropharmacol 17(3):214–231
21. Pottoo FH, Tabassum N, Javed MN, Nigar S, Rasheed R, Khan A, Barkat MA, Alam MS, Maqbool A, Ansari MA, Barreto GE (2019) The synergistic effect of raloxifene, fluoxetine, and bromocriptine protects against pilocarpine-induced status epilepticus and temporal lobe epilepsy. Mol Neurobiol 56(2):1233–1247
22. Ferreira VF, Pinto AC (2010) A fitoterapia no mundo atual. Quím Nova 33(9):1829–1829
23. Phillipson JD (2001) Phytochemistry and medicinal plants. Phytochemistry 56(3):237–243
24. Mara Mainardes R et al (2006) Liposomes and micro/nanoparticles as colloidal carriers for nasal drug delivery. Curr Drug Deliv 3(3):275–285
25. Grill AE et al (2009) A review of select recent patents on novel nanocarriers. Recent Pat Drug Deliv Formul 3(2):137–142
26. Chorilli M et al (2007) Aspectos gerais em sistemas transdérmicos de liberação de fármacos. Rev Bras Farm 88(1):7–13
27. Bonifacio BV et al (2014) Nanotechnology-based drug delivery systems and herbal medicines: a review. Int J Nanomedicine 9:1
28. Harshita A et al (2019) Nanopaclitaxel therapy: an evidence based review on the battle for next-generation formulation challenges. Nanomedicine 14(10):1323–1341
29. Sofias AM et al (2017) The battle of "nano" paclitaxel. Adv Drug Deliv Rev 122:20–30
30. Bernabeu E et al (2017) Paclitaxel: what has been done and the challenges remain ahead. Int J Pharm 526(1-2):474–495
31. Barkat MA et al (2019) Paclitaxel-loaded nanolipidic carriers with improved oral bioavailability and anticancer activity against human liver carcinoma. AAPS PharmSciTech 20(2):87
32. Kundu P, Mohanty C, Sahoo SK (2012) Antiglioma activity of curcumin-loaded lipid nanoparticles and its enhanced bioavailability in brain tissue for effective glioblastoma therapy. Acta Biomater 8:2670–2687
33. Li Z et al (2012) Antitumor activity of celastrol nanoparticles in a xenograft retinoblastoma tumor model. Int J Nanomedicine 7:2389

34. Shi F et al (2012) Preparation and characterization of solid lipid nanoparticles loaded with frankincense and myrrh oil. Int J Nanomedicine 7:2033
35. Deepa V et al (2012) Nanoemulsified ethanolic extract of Pyllanthus amarus Schum & Thonn ameliorates CCl 4 induced hepatotoxicity in Wistar rats. Indian J Exp Biol 50(11):785–794
36. Han L et al (2012) Co-encapsulation and sustained-release of four components in ginkgo terpenes from injectable PELGE nanoparticles. Fitoterapia 83(4):721–731
37. Chang L-C et al (2011) Preparation, characterization and cytotoxicity evaluation of tanshinone IIA nanoemulsions. J Biomed Nanotechnol 7(4):558–567
38. Hu L et al (2010) Preparation and enhanced oral bioavailability of cryptotanshinone-loaded solid lipid nanoparticles. AAPS PharmSciTech 11(2):582–587
39. Sutthanut K et al (2009) Solid lipid nanoparticles for topical administration of Kaempferia parviflora extracts. J Biomed Nanotechnol 5(2):224–232
40. Yadav D et al (2011) Novel approach: herbal remedies and natural products in pharmaceutical science as nano drug delivery systems. Int J Pharm Technol 3(3):3092–3116
41. Siddique YH et al (2013) Synthesis of alginate-curcumin nanocomposite and its protective role in transgenic Drosophila model of Parkinson's disease. ISRN Pharmacol 2013:794582
42. Wang W et al (2012) Enhanced bioavailability and efficiency of curcumin for the treatment of asthma by its formulation in solid lipid nanoparticles. Int J Nanomedicine 7:3667
43. Nair KL et al (2012) Purely aqueous PLGA nanoparticulate formulations of curcumin exhibit enhanced anticancer activity with dependence on the combination of the carrier. Int J Pharm 425 (1-2):44–52
44. Doggui S et al (2012) Neuronal uptake and neuroprotective effect of curcumin-loaded PLGA nanoparticles on the human SK-N-SH cell line. J Alzheimers Dis 30(2):377–392
45. Tsai Y-M et al (2011) Optimised nano-formulation on the bioavailability of hydrophobic polyphenol, curcumin, in freely-moving rats. Food Chem 127(3):918–925
46. Chang C-Z et al (2015) Curcumin, encapsulated in nano-sized PLGA, down-regulates nuclear factor κB (p65) and subarachnoid hemorrhage induced early brain injury in a rat model. Brain Res 1608:215–224
47. Tsai W-C et al (2012) Constructing liposomal nanovesicles of ginseng extract against hydrogen peroxide-induced oxidative damage to L929 cells. Food Chem 132(2):744–751
48. Mathiyalagan R et al (2014) Ginsenoside compound K-bearing glycol chitosan conjugates: synthesis, physicochemical characterization, and in vitro biological studies. Carbohydr Polym 112:359–366
49. Frozza RL et al (2010) Characterization of trans-resveratrol-loaded lipid-core nanocapsules and tissue distribution studies in rats. J Biomed Nanotechnol 6(6):694–703
50. Lu X et al (2009) Resveratrol-loaded polymeric micelles protect cells from Aβ-induced oxidative stress. Int J Pharm 375(1-2):89–96
51. Lee C-W et al (2012) Resveratrol nanoparticle system improves dissolution properties and enhances the hepatoprotective effect of resveratrol through antioxidant and anti-inflammatory pathways. J Agric Food Chem 60(18):4662–4671
52. Guo W et al (2013) Transferrin modified PEG-PLA-resveratrol conjugates: in vitro and in vivo studies for glioma. Eur J Pharmacol 718(1-3):41–47
53. Jung K-H et al (2015) Resveratrol-loaded polymeric nanoparticles suppress glucose metabolism and tumor growth in vitro and in vivo. Int J Pharm 478(1):251–257
54. Dube A, Nicolazzo JA, Larson I (2010) Chitosan nanoparticles enhance the intestinal absorption of the green tea catechins (+)-catechin and (−)-epigallocatechin gallate. Eur J Pharm Sci 41 (2):219–225
55. Liang J et al (2014) Cytotoxicity and apoptotic effects of tea polyphenol-loaded chitosan nanoparticles on human hepatoma HepG2 cells. Mater Sci Eng 36:7–13
56. Zou L-Q et al (2014) Improved in vitro digestion stability of (−)-epigallocatechin gallate through nanoliposome encapsulation. Food Res Int 64:492–499
57. Chen C-C et al (2014) Improving anticancer efficacy of (−)-epigallocatechin-3-gallate gold nanoparticles in murine B16F10 melanoma cells. Drug Des Devel Ther 8:459

58. Hong Z et al (2014) Improving the effectiveness of (−)-epigallocatechin gallate (EGCG) against rabbit atherosclerosis by EGCG-loaded nanoparticles prepared from chitosan and polyaspartic acid. J Agric Food Chem 62(52):12603–12609
59. Dhawan S, Kapil R, Singh B (2011) Formulation development and systematic optimization of solid lipid nanoparticles of quercetin for improved brain delivery. J Pharm Pharmacol 63 (3):342–351
60. Li H et al (2009) Enhancement of gastrointestinal absorption of quercetin by solid lipid nanoparticles. J Control Release 133(3):238–244
61. Karadag A, Ozcelik B, Huang Q (2014) Quercetin nanosuspensions produced by high-pressure homogenization. J Agric Food Chem 62(8):1852–1859
62. Wang G et al (2012) Effects of quercetin nanoliposomes on C6 glioma cells through induction of type III programmed cell death. Int J Nanomedicine 7:271
63. Okonogi S, Riangjanapatee P (2015) Physicochemical characterization of lycopene-loaded nanostructured lipid carrier formulations for topical administration. Int J Pharm 478(2):726–735
64. Ha TVA et al (2015) Antioxidant activity and bioaccessibility of size-different nanoemulsions for lycopene-enriched tomato extract. Food Chem 178:115–121
65. de Oliveira CC et al (2006) Canova, a Brazilian medical formulation, alters oxidative metabolism of mice macrophages. J Infect 52(6):420–432
66. Ramachandran C et al (2007) Investigation of cytokine expression in human leukocyte cultures with two immune-modulatory homeopathic preparations. J Altern Complementary Med 13 (4):403–408
67. Lussignoli S et al (1999) Effect of Traumeel S, a homeopathic formulation, on blood-induced inflammation in rats. Complement Ther Med 7(4):225–230
68. Oberbaum M et al (2005) Antiviral activity of Engystol®: an in vitro analysis. J Altern Complement Med 11(5):855–862
69. Hostanska K et al (2012) A homeopathic remedy from arnica, marigold, St. John's wort and comfrey accelerates in vitro wound scratch closure of NIH 3T3 fibroblasts. BMC Complement Altern Med 12(1):1–10
70. Bellavite P et al (2012) Testing homeopathy in mouse emotional response models: pooled data analysis of two series of studies. Evid Based Complementary Altern Med 2012:954374
71. Sunila ES et al (2009) Dynamized preparations in cell culture. Evid Based Complementary Altern Med 6(2):257–263
72. Pathak S et al (2007) Supportive evidence for the anticancerous potential of alternative medicine against hepatocarcinogenesis in mice. Complementary Med Res 14(3):148–156
73. Ghosh S et al (2013) Homeopathic mother tincture of Phytolacca decandra induces apoptosis in skin melanoma cells by activating caspase-mediated signaling via reactive oxygen species elevation. J Integr Med 11(2):116–124
74. Ruckmani A et al (2012) Anxiolytic effect of homeopathic preparation of Pulsatilla nigricans in Swiss albino mice. Homeopathy 101(3):171–174
75. Malsch NH (2005) Biomedical nanotechnology. CRC Press, Boca Raton
76. Resnik DB, Tinkle SS (2007) Ethical issues in clinical trials involving nanomedicine. Contemp Clin Trials 28(4):433–441
77. Resnik DB, Tinkle SS (2007) Ethics in nanomedicine. Nanomedicine 2(3):345–350
78. Gordijn B (2005) Nanoethics: from utopian dreams and apocalyptic nightmares towards a more balanced view. Sci Eng Ethics 11(4):521–533
79. Dupuy J-P (2007) Complexity and uncertainty: a prudential approach to nanotechnology. In: Nanoethics: the ethical and social implications of nanotechnology. Wiley, Hoboken, pp 119–131
80. Allhoff F et al (2010) Ethics of human enhancement: 25 questions & answers. Stud Ethics Law Technol 4:1
81. Mohsin J, Zahra B (2008) Protein nanoparticle: a unique system as drug delivery vehicle. Afr J Biotechnol 7(20):4926–4934

82. Pragati S et al (2009) Solid lipid nanoparticles: a promising drug delivery technology. Int J Pharm Sci Nanotechnol 2:509–516
83. Mukherjee S, Ray S, Thakur R (2009) Solid lipid nanoparticles: a modern formulation approach in drug delivery system. Indian J Pharm Sci 71(4):349
84. Sudarshan B et al (2009) Tamoxifen citrate loaded solid lipid nanoparticles-a novel approach in the treatment of ER+ breast cancer. Res J Pharm Dosage Forms Technol 1(2):143–149
85. Prabhu P, Patravale V (2012) The upcoming field of theranostic nanomedicine: an overview. J Biomed Nanotechnol 8(6):859–882
86. Muthu MS et al (2014) Nanotheranostics· application and further development of nanomedicine strategies for advanced theranostics. Theranostics 4(6):660
87. Mülhopt S et al (2018) Characterization of nanoparticle batch-to-batch variability. Nanomaterials 8(5):311
88. Saraf S et al (2015) Advancements and avenues in nanophytomedicines for better pharmacological responses. J Nanosci Nanotechnol 15(6):4070–4079
89. Gunasekaran T et al (2014) Nanotechnology: an effective tool for enhancing bioavailability and bioactivity of phytomedicine. Asian Pac J Trop Biomed 4:S1–S7
90. Kingston DGI (2011) Modern natural products drug discovery and its relevance to biodiversity conservation. J Nat Prod 74:496–511
91. Qureshi NA (2013) Mood disorders and complementary and alternative medicine: a literature review. Neuropsychiatr Dis Treat 9:639–658
92. Egert S (2011) Which sources of flavonoids: complex diets or dietary supplements? Adv Nutr 2:8–14
93. Kaur IP et al (2014) Issues and concerns in nanotech product development and its commercialization. J Control Release 193:51–62
94. Khorasani AA, Weaver JL, Salvador-Morales C (2014) Closing the gap: accelerating the translational process in nanomedicine by proposing standardized characterization techniques. Int J Nanomedicine 9:5729
95. Wei A, Mehtala JG, Patri AK (2012) Challenges and opportunities in the advancement of nanomedicines. J Control Release 164(2):236–246
96. Liu A et al (2008) Toxicological effects of multi-wall carbon nanotubes in rats. J Nanopart Res 10(8):1303–1307
97. Borm PJ, Kreyling W (2004) Toxicological hazards of inhaled nanoparticles—potential implications for drug delivery. J Nanosci Nanotechnol 4(5):521–531
98. Dreher KL (2004) Health and environmental impact of nanotechnology: toxicological assessment of manufactured nanoparticles. Toxicol Sci 77(1):3–5
99. Patel S, Nanda R, Sahoo S (2015) Nanotechnology in healthcare: applications and challenges. Med Chem 5(21):528–533

Nanomedicine Based Phytoformulation in Disease Diagnosis and Treatment

Alok Sharma, Kuldeep Singh Yadav, Faheem Hyder Pottoo, Vineet Kumar Rai, and Md. Abul Barkat

Abstract

In the last several decades, numerous investigations have been done for the development of the novel carrier systems for the delivery of phytoconstituents. Several nano-phytomedicines, for instances, nanoparticles, nanocomposites, lipid carriers, phytosomes, nanoemulsions, microcapsules, ethosomes, etc., have been developed to improve effectiveness of phytoconstituents in the management of an array of indications. These investigations are made when scientific reports question the virtue of phytomedicine. Low lipid solubility, poor stability, large size moiety, and undesired metabolism in gut are some of the critical issues of phytomedicine that have been highlighted most. Novel drug delivery systems are gradually being recognized as the most efficient carriers that have ability to overcome safety and efficacy related issues of the phytoactives. In addition, these are helping design therapeutically effective nano-phytomedicines. In addition, disease diagnosis and molecular imaging have become the most significant application of nano-phytomedicine now. Use of these carriers in current preclinical research or clinical practices as a diagnosis tool, magnetic resonance imaging, computed tomography, ultrasound, optical imaging and photoacoustic imaging, positron emission tomography, and single photon emission computed tomography, has gained momentum and is now taking a revolutionary shift as potential

A. Sharma · V. K. Rai (✉)
ISF College of Pharmacy, Moga, Punjab, India

K. S. Yadav
CSIR-Central Institute of Medicinal and Aromatic Plants, Lucknow, India

F. H. Pottoo
Department of Pharmacology, College of Clinical Pharmacy, Imam Abdulrahman Bin Faisal University, Dammam, Saudi Arabia

M. A. Barkat
Department of Pharmaceutics, College of Pharmacy, University of Hafr Al-Batin, Hafr Al-Batin, Saudi Arabia

S. Beg et al. (eds.), *Nanophytomedicine*,
https://doi.org/10.1007/978-981-15-4909-0_11

imaging modalities. Furthermore, they have shown enormous potential in molecular imaging of, drugs permeability and retention through biological membrane, absorption mechanism, their targeted delivery at cellular, tissue and organ level, tumor formation, angiogenesis, mononuclear phagocytic systems, etc. The present chapter highlights the status of the developed nano-phytomedicines, their clinical applications in the disease treatment and diagnosis.

Keywords

Nanoparticles · Nanocomposites · Lipid carriers · Phytosomes · Nanoemulsions · Microcapsules · Ethosomes · Phytoconstituents

1 Introduction

Owing to the low solubility, the phytochemicals are not being considered suitably fit for direct utilization in the development of pharmaceutical dosage forms [1]. As crude extracts contain an array of pharmaceutical compounds, some of the active ingredients present in those extract may give wrong information while screened by high-throughput technology. Lipophilicity and out of place size offer poor absorption and bioavailability of the herbal drugs [2]. These are also associated with structural instability in biological environment [1]. Biotransformation of some of the plant extracts occurs in gastric secretions and sometime gets destroyed during gastric emptying following oral administration [3]. The uses of nanotechnology for the management, identification, monitoring and managing biological system have recently named as nanomedicine. In herbal field, nano-based herbal formulation has a great potential [4–6]. In nano-phytomedicine research, the development of nanocarriers offered many advantages for better utilization of the herbal medicines, such as enhancement of bioavailability, improvement of chemical and physical stability, and avoiding drug degradation, and many more [7]. Nano-phytomedicines are developed primarily to modulate these physicochemical properties of the herbal drugs [8, 9]. Subsequently, synergistic drug combinations offered significant improvement in clinical management of various indications i.e. from infection to neuronal disorders [10–13]. Disease diagnosis is very novel area, i.e. meant for diagnosis of disease or highlighting various steps involved in the diseases treatment. For the same purposes, various techniques including different kinds of contrast agents are used [14]. In order to make this process safe, cost effective, and easier, various kinds of contrast agents have been investigated and focus of the research is now sifted to novel carrier based contrast agents [8, 9]. Use of these carriers in current preclinical research or clinical practices as a diagnosis tool, magnetic resonance imaging, computed tomography, ultrasound, optical imaging and photoacoustic imaging, positron emission tomography and single photon emission computed tomography has gained momentum and is taking a revolutionary shift in potential imaging modalities. Not only disease diagnosis, molecular imaging techniques are being used to detect drugs permeability through biological

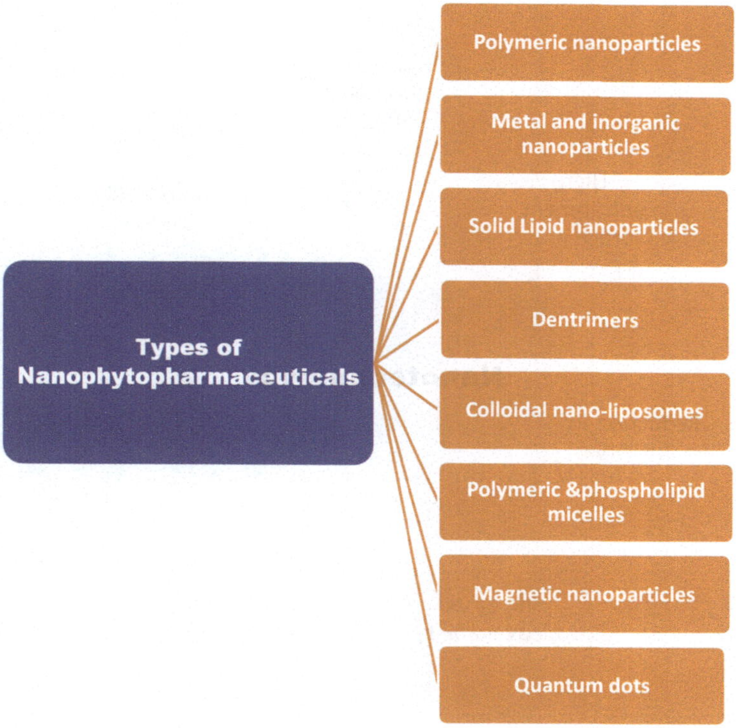

Fig. 1 Different types of nanocarriers

membrane, tissue and organ retention, targeted delivery of drugs at cellular, tissue and organ level, tumor formation, angiogenesis, mononuclear phagocytic systems, etc. [15]. Application of nano-phytomedicine is now widen from, solubility enhancement to bioavailability enhancement, modulation of pharmacokinetic profile to pharmacodynamic profile, curing of diseases to treatment of diseases and disease diagnosis to molecular imaging and now we are on the stage on which a paradigm shift is expected in this field to widen its industrial scalability. Following section includes the potential of nano-phytomedicine for improved therapeutic effectiveness of phytomedicine in disease treatment and diagnosis.

2 Nano-Phytomedicine in the Management of Diseases

Nano-phytomedicine is the collection of word nanocarriers and phytomedicine. In this science, physicochemical properties of the phytomolecules are modified and outcome of this modification is measured in terms of improved safety and efficacy [16]. Figure 1 highlights different types of nanocarriers that have been developed using phytomedicines. Polymeric, metal, solid lipid, colloidal, magnetic

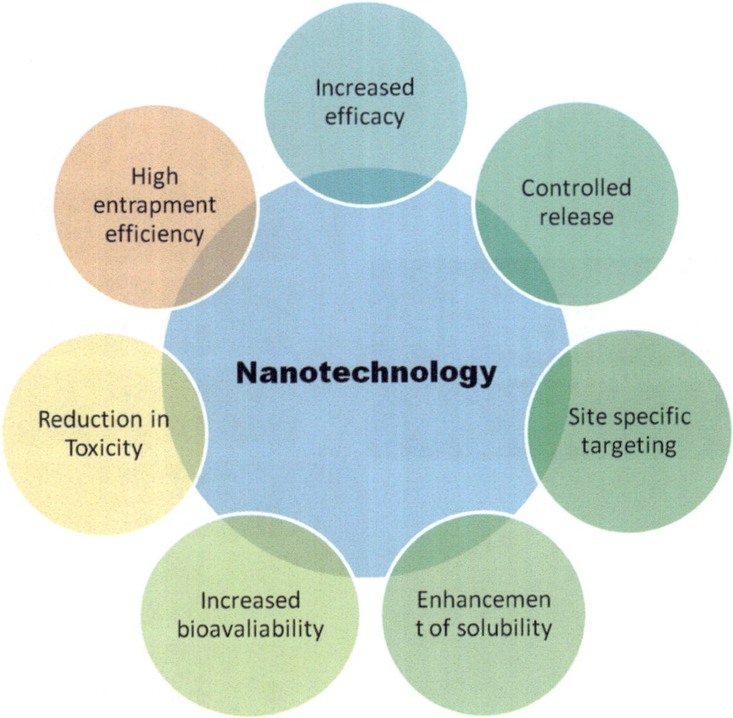

Fig. 2 Effect and role of nano-phytopharmaceuticals in novel drug delivery system

nanoparticles, dendrimers, quantum dots, nano emulsions, liposomes are some of them. Figure 2 represents major advantages that can be obtained when nanocarriers of phytoconstituents are prepared. Improved efficacy is expected as the final outcome of physicochemical modification of the phytoactives by using these carriers. The treatment of diseases using nano-phytomedicine has reasonably been done. Recent advancement of the treatment of diseases by phytomedicine has gained momentum. Most significant therapeutic applications of nano-phytomedicine that have been investigated till date are cancer treatment [16], hepato-protection [3], retardation of oxidation, treatment of cardiovascular disorders, neurological disorders, autoimmune/rheumatological disorders, inflammation [17], avoidance of nutritional deficiency, etc. (Fig. 3).

Nanocarriers have been used as some of the most effective measures for the enhancement of therapeutic effectiveness of the drug as well as to ensure drug delivery at the specific site. In contrast to the conventional drug delivery system, nanocarriers have higher potential in improving solubility, stability, permeability, and bioavailability with other benefits such as longer half-life, site-specific drug delivery, and reduced side effects [3]. In recent years, the concept of nano-phytomedicine has gained much attention. The ideas and amelioration mechanisms of the nanomedical devices for the delivery of phytomedicines as therapeutic agent

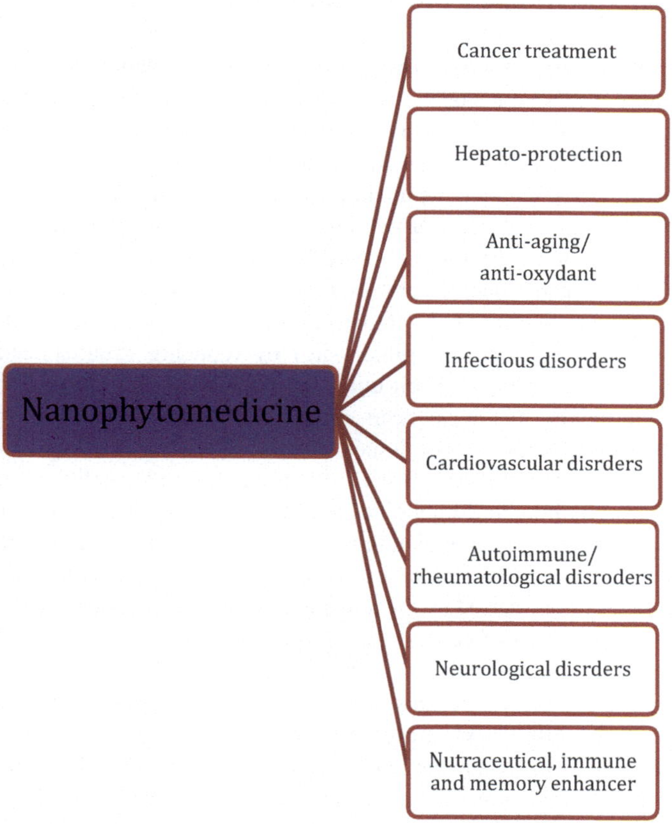

Fig. 3 Nano-phytomedicine to improve the therapeutic effectiveness of phytomedicine in the management of various diseases and disorders

are increasingly being investigated. These modalities are useful in the therapy of brain, lung, breast, and colon and rectal cancers. These nanocarrier-based systems for phytomedicines are majorly divided into three categories such as polymeric or solid, lipid, and colloidal or liquid systems. Nanocarriers have been utilized for the delivery of so many phytomedicines such as curcumin, paclitaxel, wogonin, quercetin, catechins, green tea, artemisinin, berberine, camptothecin, curcuminoids, taxel, zedoary turmeric oil, docetaxel in chemotherapy, hepato-protection, rheumatological disorders, cardiovascular diseases, and many more. Nano-phytomedicine has also been utilized to make treatment safe to use. One of the most significant applications of this kind is reported by Mishra et al. in 2015 that indicates the potential of encapsulation of in reducing the skin irritation potential of *Mentha spicata* L. var. viridis oil along with the improved local efficacy of this oil in fungal infections [18].

2.1 Nano-phytomedicine as Liposomes

The prospects of novel drug development techniques considering phytopharmaceuticals can suggest new directions and endless frontier as novel herbal drug regimen [19]. Liposomes are microparticular colloidal lipid carriers usually 0.05–5.0 μm in diameter. These are spontaneously formed when certain lipids are hydrated in an aqueous medium. Under the interaction with water, polar lipids form self-organized nanocarriers [20]. Liposomes are spherical particles that encapsulate a portion of the solvent where they are free to diffuse (float) into their interior. They can carry one, several or more concentric membranes. Liposomal-based drug delivery systems have immense potential in increasing the therapeutic index of anti-cancer agents (at tumor site) by reducing drug concentration in systemic circulation (Table 1). Their utility in enhancing permeability and retention are the major factors that have been considered during targeting of the phyto-constituents to the specific cells or tissues. The major benefits of using lipoid carriers comprise (1) biocompatibility, (2) ease of preparation, (3) ability of loading all hydrophilic, amphiphilic and lipophilic molecules, and (4) ability to alter the pharmacokinetic parameters by modulating the physicochemical properties of the drugs [21]. Liposomes can act as effective drug delivery systems for oncology therapy as it reduces the side effects caused by the non-selective nature of cancer management by increasing the specificity of cancer cells. Liposomes has the capacity to encourage passive targeting of cancer cells [20]. Tumor-specific cells have increased permeability and retention effects because larger molecules can enter their interior. This property of cancer cells allows liposomes loaded with anti-cancer drugs to enter the cancer cells, but also enter the endothelial cell wall. Therefore, drug-loaded liposomes are directed only to tumor cells and not to healthy cells, a property known as passive targeting. Various herbal liposome formulations have been explored for different therapeutic purposes, which are summarized in Table 1. Silymarin marianum is basically extracted from fruits and mature seeds of the plant milk thistle. Their main components are isomeric flavonolignans, namely silibinin, silychristin, and silidianin [22]. Silybin is considered to be a small, highly functional molecule, however, due to the low aqueous solubility, poor permeability, high metabolism and rapid systemic elimination, the clinical application of silybin show some limitations [23]. The nanotechnology-based drug delivery systems exploit enormous potential of milk thistle by improving its water solubility and bioavailability. Incorporation of silymarin into liposome formulations enhances its bioavailability.

2.2 Nano-phytomedicine as Nanoparticles

Over the past several years, the green chemistry of herbal medicines has gained attention worldwide some of them are shown in Table 2. The primary aim behind preparing nanoparticle of therapeutic purpose is that because it is a delivery arrangement to modulate particle size, surface properties, and drug release kinetics of

Table 1 List of liposomes developed for the delivery of phytoactives

S. No.	Nano-phytomedicine	Plant source	Indication	Properties	References
1	Catechins phytosomes	*Camellia sinensis*	Antioxidant, anti-cancer, neutraceutical, immune and memory enhancer	Increased (-)-epigallocatechin gallate accumulation in basal cell carcinomas (BCCs)	[24]
2	Paclitaxel liposomes	Pacific yew tree *(Taxus brevifolia)*	Anti-cancer	Lipidic carrier improved stability of the drug	[25, 26]
3	Glycyrrhetinic acid (GA)-modified Wogonin liposomes	*Glycyrrhiza glabra*	Liver cancer	Improved biodistribution, drug accumulation at the tumor site and therapeutic efficacy due to the increased receptor-mediated uptake of liposomes by liver-targeted cells	[27]
4	Curcumin liposomes	*Curcuma longa*	Antioxidant, anti-cancer, anti-cancer	Improved oral bioavailability by stabilizing curcumin at physiological pH, improving aqueous solubility	[28]
5	Breviscapine liposomes	*Erigeron breviscapus.* (Vant.)	Cardiovascular disorders	Increased biodistribution of breviscapine in heart	[29]
6	Quercetin and rutin Liposomes	*Carpobrotus edulis*	Antioxidant, anti-cancer and in cardiovascular disorders	Controlled release, site-specific drug delivery to the cancer site	[30, 31]
7	Usnea acid liposomes	*Usnea spp.*	Anti-microbial	USFDA reports liver toxicity due to the presence of usnic acid rich dietary supplements. Liposomes reduced the toxicity	[32]
8	Colchicine liposomes	*Colchicum autumnale*	Inflammation rheumatologic/ autoimmune disorders	PEGylated colchicine-derived prodrugs controlled colchicine release and hence improved activity	[33]
9	Silymarin liposomes	*Silybum marianum*	Hepato-protection	Improved oral bioavailability and site-specificity to the liver	[20]

Table 2 List of photoactive loaded nanoparticles for therapeutic purpose

S. No.	Formulations	Formulation properties	Method of preparation	Biological activity	References
1	Solid lipid nanoparticles of curcuminoids	Improved solubility and distribution	Microemulsion technique	Anti-cancer and antioxidant	[28]
2	Glycyrrhizinic acid nanoparticles	Improved distribution in systemic circulation	Rotary-evaporated film ultra-sonication method	Anti-inflammatory antihypertensive	[35]
3	Nanoparticles of Cuscuta chinensis	Improved oral absorption of the drug	Nanosuspension method	Hepatoprotective and antioxidant effects	[36]
4	Taxel nanoparticles	Improved therapeutic index of the drug and devoid of the adverse effects	Emulsion solvent evaporation method	Anti-cancer	[37]
5	Camptothecin nanoparticles	Improved site specificity and distribution	Dialysis method	Anti-cancer	[38]
6	Artemisinin nanocapsules	Improved safety and distribution	Self-assembly procedure	Anticancer	[39]
7	Berberine nanoparticles	Enhanced in vitro bioavailability	Ionic gelation method	Anticancer	[40]
8	Camptothecin nanoparticles	Increased accumulation of nanoparticles at a tumor site	Sinigrin precipitation method	Anti-cancer	[38]
9	Zedoary turmeric oil nanoparticle	Improved hydro-philicity, stabilities and bio-availabilities	Emulsification cum solidification method	Anti-microbial, hepato-protectant, anti-cancer	[41]
10	Triptolide nanoparticle	Reduced toxicity, improved aqueous solubility, site specificity at the therapeutic targets	Galactosylation of chitosan and fabrication	Inflammation rheumatologic/ autoimmune disorders	[42, 43]
11	Glycyrrhizic acid nanoparticle	Improved in vitro inflammatory activity	Double-emulsion evaporation cum solvent evaporation method	Inflammation rheumatologic/ autoimmune disorders	[35]
12	Radix salvia miltiorrhiza nanoparticles (CHF)	Increased distribution in to systemic circulation	Spray-drying technique	Against cardiovascular disorders	[44]
13	Flavonoids and lignans nanoparticles	Improved aqueous solubility of flavonoids and lignans for improving oral absorption	–	Hepato-protection	[36]
14	Silibinin loaded nanoparticles	Enhanced oral bioavailability and hepatic protection of poorly water-soluble drugs	Self-assembling of solubilized lipids	Hepato-protection	[45]

pharmaceutically or therapeutically active components to accomplish the targeted action of photoactive at the desired rate. Nanoparticles are colloidal systems with particle sizes ranging from 10 nm to 1000 nm. Nanocarriers with mean particle diameter more than 100 nm are also reported for instances, nano-sized curcuminoids, paclitaxel, and praziquantel, with a mean particle diameter of 147 nm and more than 200 nm. In recent decades, biodegradable polymer based nanoparticles have attracted the attention of the researchers considerably as proficient drug delivery carrier systems. Nanoparticles are used to target the herbal medicine to individual organ, which generally improves the selectivity, drug delivery, effectiveness, and safety, and thereby reduces the dose and increases patient compliances. Nanoparticles can be utilized to increase the herbal drug solubility and help to localize the drug in a specific site thus resulting in better efficacy. They appear to be able to deliver high concentrations of drugs to disease sites because of their unique size and high loading capacities. The delivery of the drug in small particle size enhances the entire surface area of the drugs and due to this they show the allocating quicker dissolution profile in the blood. It also shows high EPR (enhanced permeation and retention) effect, i.e., enhanced permeation through the barriers because of the small size and retention due to poor lymphatic drainage such as in tumor. They also exhibit passive targeting to the disease site of action without the addition of any particular ligand moiety. Nano-crystallization has many advantages over various conventional herbal formulations, such as increasing the solubility of the compound, reducing the dosage of the herbal drug, and improving the absorption [34] of the herbal medicines.

2.3 Nano-phytomedicine as Phytosome

The herbal research has established various scientifically validated phytoconstituents with biological activity and health benefits. Most of the active ingredients are polar or water-soluble. These water-soluble molecules are not absorbed due to large molecular size, or are poorly absorption property and this leads to poor bioavailability. Unique properties of phytophospholipid complex (Phytosome) along with their application in Novel herbal formulation have assured the researchers to deliver bioactive to their systemic targets. Various methods employed in phytosomal preparation and characterization along with the phytosomal advantages over conventional herbal extracts is described in the published literature. It advanced herbal drug technology that offers defined bioavailability of phytoconstituents over the conventional herbal formulation. Phytosomes are lipid complex between a natural moieties and soy lecithin compounds in a stoichiometric ratio. Reduced particle size, upturned rate of dissolution, and absorption claims enhanced in vivo performance of the herbal formulation and extracts [46]. Phytosome is a patented technology, which is prepared by one of the leading pharmaceutical, and nutraceutical manufacturers that incorporate standardized herbal extracts or botanical ingredients having aqueous solubility into phospholipids to produce lipid-compatible molecular complexes known as plant bodies. Structural elucidation reveals that the chemical interaction of phospholipids and herbal bioactive involves the formation of hydrogen bonds amid the polar side front of phospholipids and the polar functional group producing a specific pattern. These show chemical reaction in solvents and assume a micelles shape network in water appearing liposomal-like cellular configuration in which the active polar moiety is docked to phospholipids behaving as integral part of the cell membrane [47]. There are numerous factors, which govern the characteristic originality of phytosomes in physical state like size, membrane permeability, entrapment ratio, chemical constitution, and the quantity and purity of the precursor starting chemical ingredients. Phytosome has evolved as an important revolution for conventional herbal formulations which were not able to demonstrate a remarkable effect at pharmacological effective (Table 3). Phytosome technology has emerged as safe and promising drug delivery of option for herbal medicines and extract. The phytosome upgrade absorption of lipophobic polar herbal extract through oral route as well as topical route, which shows better bioavailability. As absorption and bioavailability of the phytoconstituent increases the bioavailability, so the requirement of dose is reduced. Phytosomes have better stability profile due to presence of chemical bonds between phospholipid and phytoconstituent. One such method is by increasing the solubility of bile to herbal origin phytoconstituents, phytosomes enhance the liver targeting [48]. The phytosome possess better drug entrapment efficient such as phosphatidylcholine used in the preparation of phytosomes, as such it is not a carrier; but it also act as hepatoprotective, hence giving the synergistic effect when it is employed as hepatoprotective substance or drug [49]. Phytosomes can also be used for systemic targeting because of its ability to better its transition from hydrophilic condition to fat soluble environment of enterocyte cell membrane.

2.4 Nano-phytomedicine as Nano and Microemulsion

Microemulsions refer to clear, thermodynamically stable, consisting of an isotropic mixture of oil and water, stabilized by surfactants and sub-surfactants. Typically, the emulsion consists of droplet-type dispersions, either oil in a water type or water in oil type. The size ranges of emulsion from 0.1 to 100 μm, a microemulsion 10–100 nm, a sub-microemulsion 100–600 nm. There are two types of microemulsion, one is O/W and the second is W/O microemulsion. For preparation of O/W microemulsion, we start with w/o emulsion using a low hydrophilic–lipophilic balance (HLB) number surfactant [1, 2]. To this emulsion, an aqueous solution of high HLB number surfactant is added while stirring at a certain amount of addition, a "gel" phase is produced and further addition of surfactant solution, an inversion into O/W emulsion take place. For W/O microemulsion, one start with O/W emulsion stabilized with an ionic or nonionic surfactant. This emulsion is titrated with a co-surfactant and the emulsion passes through a gel phase, after which further addition of co-surfactant result in the production of W/O microemulsion [59]. As a novel drug delivery carrier system, the emulsion is distributed in the body in a targeted manner due to its affinity for lymph. The drug release is sustained because the drug is present in the inner phase. The size of the emulsion particles affects the target distribution. Additionally,

Table 3 List of Phytosomes developed encapsulating phytomedicines

S. No.	Formulation	Plant source	Use/indication and formulation characteristics	References
1	Green tea phytosome	*Camellia sinensis*	Antioxidant, cardioprotective, and protection against cancer	[24]
2	Ginkgo biloba phytosome	*Gingko biloba* (Ginsenosides)	Protects brain and vascular lining, adaptogenic	[50]
3	Centella phytosome	*Centella asiatica* (Triterpine)	Cicatrizing, trophodermic, anti-inflammatory effect	[51]
4	Silymarin phytosome	*Silybum marianum* (Silymarin)	Anti-hepatotoxic	[52, 53]
5	Visnadine phytosome	*Ammi visnaga* (Visnadine)	Circulation improver	[54]
6	Catechins phytosomes	*Camellia sinensis*	Antioxidant, anti-cancer, neutraceutical, immune and memory enhancer	[24]
7	Curcumin phytosome	*Curcuma longa*	Improved oral bioavailability by stabilizing curcumin at physiological pH and hence improved antioxidant, anti-cancer, anti-cancer properties	[28]
8	Naringenin phytosomes	*Lycopersicum esculentum*	Improved oral bioavailability of naringenin and enhanced hepatoprotective effects. Improved antioxidant and hepato-protection	[55]
9	Ginkgo biloba phytosome	*Ginkgo biloba*	Improved oral absorption and hence antioxidant activity of the drug. Improved antioxidant effect in neurological disorders	[56]
10	Grape seed phytosome	*Vitis vinifera*	Improved oral absorption and distribution. Antioxidant effect	[57]
11	Hawthorn and ginseng phytosome	*Panax ginseng*	Greater absorption of phytoactives as compared to that of its extracts. Neutraceutical, immune and memory enhancer	[58]

the sustained release herbal emulsions can also enhance the stability of the hydrolyzed substance, and improve the penetration of the drug into the skin and mucous. Emulsion containing herbal ingredient is able to increase the stability of the developed systems and increases the penetrability of drug into different skin layers mucous membranes [41]. So far, some herbs like camptothecin, brucea oil, coixenolide oil, and zedoary oil have been developed in the form of herbal emulsions (Table 4). Furthermore, they can be administered through different routes such as transdermal, parenteral, pulmonary, and ocular. Yin et al. developed Docetaxel microemulsion to improve its oral bioavailability [60]. Similarly, Wang et al. made an attempt to enhance anti-inflammatory activity of curcumin by formulating

Table 4 List nano-emulsified phytomedicines of therapeutic use

S. No.	Nano-phytomedicine	Formulation characteristics	Indication	References
1	Triptolide nanomulsion	Improved percutaneous delivery to avoid systemic toxicity of the drug	Inflammation rheumatologic/ autoimmune disorders	[64]
2	Artemisinin nanocapsules	Controlled release of the drug	Anti-cancer	[39]
3	Berberine nanoemulsion	Improved oral bioavailability by improving aqueous solubility and dissolution rate of the drug in addition to the improved permeability through caco-2 cell monolayers	Anti-cancer	[40]
4	Docetaxel submicron emulsion	Improved aqueous solubility and reduced toxicity	Anti-cancer	[65]
5	SNEDDS of zedoary essential oil	Improved oral distribution of the oil	Hepato-protection	[66]
6	Silybin nanoemulsion and phytosomes	Improved oral bioavailability by improving Silybin stability at different temperatures	Antioxidant and hepato-protectant	[53, 67]

it into nanoemulsion [61]. Megha et al. formulated nano-encapsulated poly-herbal ointment for anti-inflammation [62]. Zhaoa et al. developed self-nanoemulsifying drug delivery system (SNEDDS) for oral delivery of Zedoary essential oil [63]. Elemenum emulsion is a new type of anticancer drug with no damage to the heart and liver. The enamel emulsion significantly inhibited the proliferation and differentiation of A549 in vitro.

2.5 Nano-phytomedicine as Transferosomes, Ethosomes and Herbosomes

Transferosomes refer as artificial vesicle, which are developed to show the properties of a cell vesicle and appropriate for site-specific drug delivery. It can be used as a pharmaceutical carrier for an array of low molecular weight molecules, peptides, proteins, and herbal and bio-actives. They are phospholipid bilayer vesicular structures proposed to apply through transdermal route. Transferosomes and ethosomes increase skin penetration through the stratum corneum barrier but their action is not similar. Transferosomes act by exploiting hydration and osmotic pressure of the skin, whereas ethosomes has high content (20–45%) of ethanol. Due to the presence of chemical based penetration enhancers, ethosomes usually

Table 5 List of phytoactives loaded ethosomes

S. No.	Formulations	Active ingredients	Biological activity	Method of preparation	References
1	Glycyrrhiza glabra ammonium glycyrrhi zinate ethosomes	Glycrrhizic acid	Anti-inflammatory	Solvent dispersion method	[71]
2	Tripterygium wilfordi triptolide ethosomes	Diterpene triepoxide	Anti-inflammatory	Filming rehydration method and ultrasonic method	[72]
3	Podophyllum hexandrum ethosomes	Podophyllotoxin	Purgative, antirheumatic, antiviral and antitumor	Solvent dispersion method	[73]
4	Sesbania grandiflora sesbania ethosome	Leucocyanidin and cyaniding	Anti-microbial	Solvent dispersion method	[74]

disrupts the membrane barrier reversibly and thereby enhances the permeability [68]. Ethanosome carriers can provide effective intracellular delivery of hydrophilic and lipophilic drugs, increased transdermal absorption of matrine, and increased anti-inflammatory herbs (Table 5). Transferosomes are used for the delivery of the various phytoactives in different upper skin layers, however, for deeper penetration into the skin layers and for systemic delivery it is not a good choice. There are enormous products where the herbal constituent and their bioactive mediated treatment are enhanced by developing these drug delivery carriers.

The therapeutic activity of any herbal medicine and their extract is the based on effectiveness of bioactive compound. Herbosomes are newly introduced terms with improved bioavailability and solubility of traditional plant extracts. The term "Herbo" means a plant or herb and "some" means a cell. Herbosomes technology is one of such systems incorporating phospholipids into standardized active ingredient of herbal extracts in hydrophilic flavonoids and other similar compounds through the skin or gastrointestinal tract. Phospholipids are the main building blocks and are one of the major components of cellular membranes. Herbosome enhance the bioavailability of bioactive or phyto-constituents, as they can cross the lipid bio-membranes easily, and their active components such as herbal extracts loaded in them are confined from destruction by digestive secretions [69]. Herbosomes are used in the treatment of various diseases like liver disease and heart disease. It is also used as an anti-inflammatory, lipolytic, vasokinetic, anti-edema, cicatrizing, trophodermic, neutraceutical immunomodulator, antioxidant for skin and liver, cardioprotective, anti-wrinkles, and UV protectant. The fruit of milk thistle plant contains a flavonoid known for hepatoprotective effect. Silymarin has been shown to have a positive effect in treating liver diseases of various kinds, including hepatitis, cirrhosis, fatty filtration of liver, and inflammation of bile duct. Silybin protects the liver by conserving glutathione into the parenchyma cells. While parenchyma cell (PC) helps repair and replace the cell membrane. These constituents likely offer the synergistic benefit of sparing liver cell from destruction. Green tea has got several long-term beneficial activities such as antioxidant, anticarcinogenic, antimutagenic, antiatherosclerotic, hypocholesterolemic, cardioprotective, and antibacterial effect. Despite such potential action green tea polyphenols have very poor oral bioavailability from conventional extracts. The complexion of green tea polyphenols with phospholipids strongly improves their poor oral bioavailability. Grape seed phytosome composed of oligomeric polyphenols of varying molecular size, complexed with phospholipids. The main properties of procyanidin flavonoids of grape seed are increase in total antioxidant capacity and stimulation of physiological antioxidant defences of plasma and blood through a network of mechanisms that extend beyond their great antioxidant potency [70]. Therefore, with the help of herbosome preparations, the amount of standardized herbal extracts phytoconstituents administered in body through several routes are required in lower amounts for significant therapeutic activity.

2.6 Nano-phytomedicine as Microspheres

Microspheres are defined as monolithic sphere or therapeutic agent distributed throughout the matrix either as a molecular dispersion of particles or can be defined as structure made up of continuous phase of one or more miscible polymers in which the drug particles are dispersed at the molecular or macroscopic level. Microspheres are spherical shaped particles with size range of 1–300 μm. Drugs are loaded in microspheres and are delivered to the specific site in specific dose. Their oral administration is advantageous owing to their oral and parenteral applications. Microspheres provide constant and prolonged therapeutic effect. Reduces the dosing frequency and thereby improve the patient compliance. They could be injected into the body due to the spherical shape and smaller size. Better drug utilization will improve the bioavailability and reduce the incidence or intensity of adverse effects. Microsphere morphology allows a controllable variability in degradation and drug release. Microspheres can be personalized as requirement profiles and based on site-specific drug delivery. Delivering drugs with the help of biodegradable microsphere of natural polymers has many advantages over the conventional drug delivery carriers or systems [75]. The herbal actives such as rutin, camptothecin, zedoary oil, tetrandrine, quercetine, and cynara scolymus extracts have been delivered by micro-particles. These polymers based delivery systems avoid the necessity of multiple dosing by presenting a way to offer controlled release for longer duration.

3 Nano-phytomedicine in Disease Diagnosis and Molecular Imaging

Early diagnosis of the life threatening disease means the discrepancy between life and death. It has always been difficult to detect or diagnose diseases. Various investigations have been done till date to improve molecular imaging of the disease and disease condition. Despite various technologies we are particularly interested in the emerging ways. Molecular imaging has widely been used to diagnose and monitor progression of diseases or their treatment. It is noninvasive method of anatomical assessment of functional and molecular information and allows the assessment of pathophysiological anomalies. Current in vivo or clinical investigations includes magnetic resonance imaging, computed tomography, ultrasound, optical imaging and photoacoustic imaging, positron emission tomography, and single photon emission computed tomography as potential imaging modalities. These modalities are dependent on various fundamental physical principles and, therefore, possess explicit advantages and disadvantages in respect of sensitivity and specificity to contrast agents, tissue contrast, spatial resolution, quantitativeness, and tissue penetration [15]. The choice of an imaging technique should not solely be dependent on the kinds of contrast agents but also on their anticipated uses. For instance, if complete body scan is essential, modalities like magnetic resonance imaging, computed tomography, positron emission tomography, and single photon emission computed tomography are suggested. Though, speedy and cost effective organ-specific assessment can remarkably be done by ultrasound. Optical and photoacoustic uses, however, are most appropriate for examining superficial lesions (e.g., skin, peripheral joints), endoscopic and intraoperative measures.

Contrast agents are the main components of any imaging technique. Significant contribution has been made to develop suitable and affordable contrast agents. However, not every contrast agent is appropriate for human use. Currently the use of nanoparticles as contrast agents has gained much attention. Various types of nanoparticles have been used as contrast agents that included polymeric nanoparticles, liposomes, ultrasmall superparamagnetic iron oxide (USPIO) nanoparticles and gold nanoparticles. List of nanomedicine developed for diseases diagnosis and molecular imaging is given in Table 6.

Application of nanocarrier-based contrast agents in angiography by arterial spin labeling MRI has become the answer of many clinical questions. As the scanning time in MRI and CT scan is very short, that may lessen the requirement for blood pool contrast agents. CT scan reliably detects pulmonary embolism, vascular damages and the integrity of aortic stent grafts, however, small gadolinium chelates has been proved as successful contrast agent as it improved magnetic resonance angiography of the coronary arteries, improved the recognition of atherosclerotic plaques and peripheral arterial disorders. In order to obtain specific imaging information, less than 5 nm and more than 1 mm size of diagnostic agents are more purposeful than that of 5- to 100-nm sized diagnostic nanoparticle for molecular imaging purposes [15]. Demand of nanocarriers in molecular imaging is highly depends on quick and site-specific contrast augmentation. Renal clearance of larger

Table 6 List of nanomedicine developed for diseases diagnosis and molecular imaging

Nanomedicines as contrast agent	Technique	References
Gadolinium-containing probes	MRI	[76]
Superparamagnetic iron oxide nanoparticles	MRI	[77]
Paramagnetic liposomes and polymeric nanocarriers	MRI	[78]
ParaCEST agents	MRI	[79]
Hyperpolarized probes	MRI	[80]
Iodine-based micelles	CT scan	[81]
Iodine liposomes	CT scan	[82]
Barium loaded nanoparticles	CT scan	[83]
Gold nanoparticles	CT scan	[84]
Bismuth nanoparticles	CT scan	[85]
Targeted and non-targeted gas-filled microbubbles	Ultrasound	[86]
Nanobubbles	Ultrasound	[87]
Near-infrared fluorochrome-labeled nanoparticles	Optical imaging	[88]
Quantum dots	Optical imaging	[89]
Fluorescent nanoparticle probes	Optical imaging	[90]
Gold nanoparticles, gold nanorods	Photoacoustic imaging	[91]
Carbon nanotubes	Photoacoustic imaging	[92]
Fluorescent/dye-loaded nanoparticles	Photoacoustic imaging	[93]
Radiolabeled gold nanoshells	Positron emission tomography	[94]
Polymeric nanoparticles	Positron emission tomography	[95]
Technetium-labeled gold nanoparticles	Single photon emission CT	[96]
Indium-labeled liposomes	Single photon emission CT	[97]
Nano- and microcolloids	Single photon emission CT	[98]

nanocarriers as contrast agent could be very high as these results in higher improvement of the biological half-lives of the contrast agent. It also leads to high degree of non-targeted retention in the extravascular cubicle or increased uptake by reticulo-endothelial systems. Before selecting nanodiagnostic agents, one should consider their actual clinical need, significant value addition over the conventional imaging techniques and their benefits above low-molecular weight diagnostic probes.

Recent advances in nanocarrier-based contrast agent development reflect positive trend. Tagging with radionuclides or contrast agents is being done by liposomes that are passively or actively aimed to the intended tissues. This is done to prolong the half-life of the radionuclides by ensuring their PEGylation. This technique conferred prolongation of circulation time for the liposomes and provide extra time to reach at the intended site. It helps increase accumulation of more contrast agent at a specific site. It increases the signal-to-background noise ratio also along with the signal intensity. Radiolabels are incorporated in liposomes by following mechanisms, i.e. encapsulation into aqueous core, entrapment in the liposomal lipid bilayer, surface chelation of the lipid bilayer, or drug loading into the core of the lipid carriers by improving transmembrane gradient. Surface chelation or remote loading methods are preferred as high signal intensity due to high encapsulation efficiency and fast loading. In an investigation, Erdogan et al. integrated a polychelating amphiphilic polymer for the attachment of gadolinium (Gd) in 2C5 mAb-targeted PEGylated lipid carriers and imaged tumors effectively in the mice even after 4th hour after injection. To increase the encapsulation efficiency in addition to the other benefits for signaling, Gd with integrated dextran was entrenched in the porous bilayer of the liposomes. Indocyanine green label in lung targeted surface-chelated lipid carrier facilitated Murata et al. to carry out noninvasive, real-time, near-infrared optical imaging of lipid carriers for pulmonary delivery. SPECT and PET techniques were employed to detect 99mTc, 111In, 123I, 18F, 68Ga, etc. [99].

Currently, molecular imaging technique is used as an important area of medical research. These techniques help improve early detection of disease and stage of the disease and enable us to guided and personalized therapy. Furthermore, it helps offer essential information on the therapeutic efficacy, by molecular imaging probes by visualizing and characterizing biological procedures at the cellular and molecular level. Current developments in nanotechnology have led to the investigation of so many nanoparticle-based formulations for diagnostic relevance. Diagnostic nanoparticles aspire to envisage pathologies and to advance the perceptive of vital physiological principles of a variety of disorders and their therapy. Clinically, though, the use of nanodiagnostics is restricted, owing to the multifaceted requirement of investigations on their pharmacokinetic properties and drug elimination kinetics. Therefore, majority of nanoparticle-based systems at present used in the clinics are suitable for therapeutic purposes. Despite fewer examples of nanocarriers as contrast agent, their application in the field of diagnostic and imaging through nano-phytomedicine cannot be undermined. Some of the most specific applications of nano-phytomedicine in molecular imaging field are given hereunder (Fig. 4).

3.1 Nano-phytomedicine for Imaging of Mononuclear Phagocytic Systems

Most nanocarriers serve as ideal entrants for the imaging of mononuclear phagocytic systems (MPS) which are majorly present in liver, spleen, and lymphatic systems. Most of the nanocarriers are engulfed by the MPS owing to the existence of higher number of macrophages in those tissues. Various nanocarrier-based imaging agents have been projected for CT, MRI, SPECT, ultrasound, OI, and PAI, however, merely Fe_2O_3 nanoparticles have so far been investigated for clinical imaging of MPS present in liver [100]. Most significant virtue of these nanocarriers is that these are readily cleared from the systemic circulation through phagocytosis by macrophages present in liver, spleen, and lymph nodes. Other than the intended use of the nanoparticles in liver imaging, imaging of phagocytosing cells in lymph nodes or in inflammatory lesions, longer biological half-lives are necessary that gives the particles ample of time for sufficient extravasation in the particular tissues

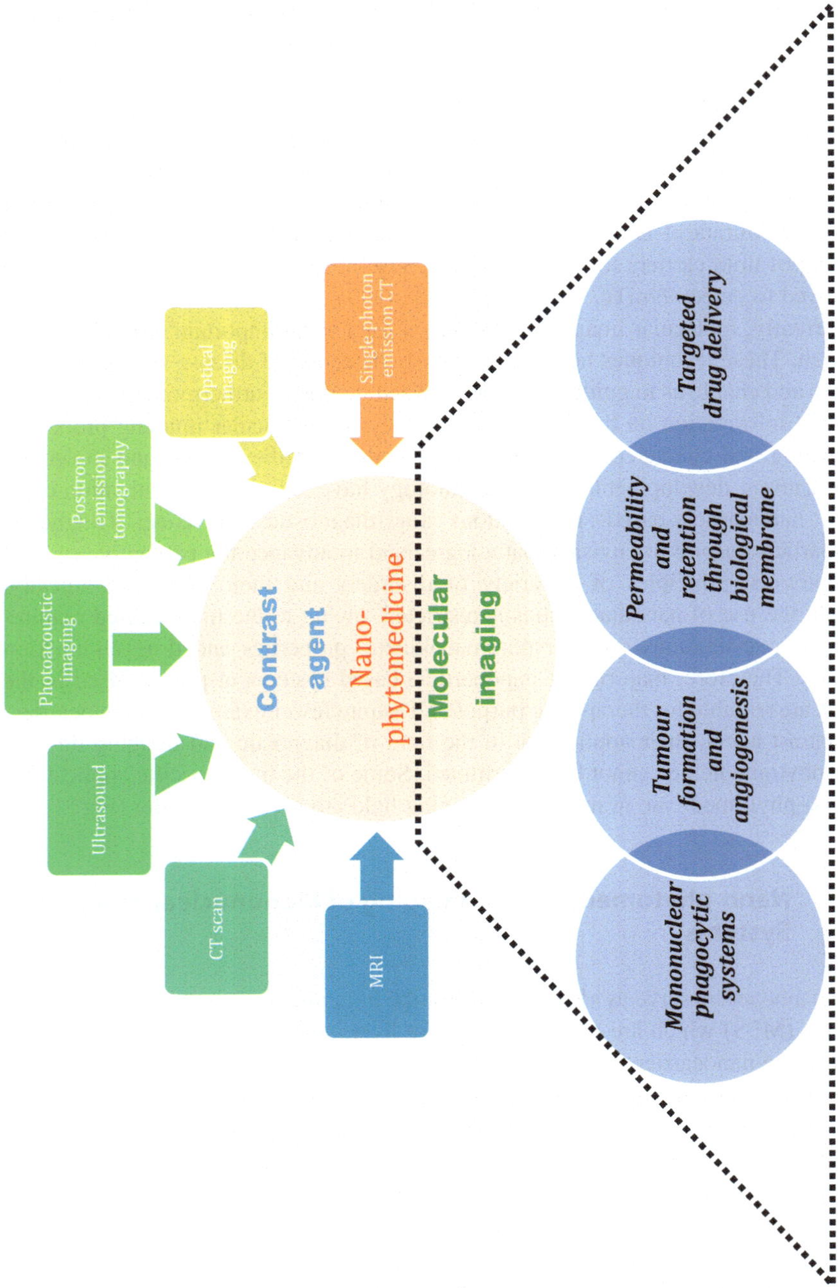

Fig. 4 Importance of nano-phytomedicine in disease diagnosis as contrast agent and molecular imaging

and to be taken up by the linked macrophages. This is achieved by making nano-sized particles with high monodisperse characteristic. Most of the reports available on detection of liver metastases, lymph node metastases and on the depiction of inflammatory lesions, such as atherosclerotic plaques by MRI using nanocarriers as contrast agent [15].

3.2 Imaging of Tumor Formation and Angiogenesis by Nano-phytomedicine

The development of tumor, angiogenesis in tumor and vascularization the major rationals behind the selction of nanocarrier-based contrast agents. Targeted nanocarriers interact with the proliferating endothelial cells and are used to detect tumor malignancy and to evaluate mechanistic alterations in tumor vascularization, like vessel maturation at the time of anti-angiogenic therapy or vascular inflammation while radiotherapy. Most of the nanocarriers based molecular imaging means aim angiogenesis because endothelial targets are easy to get to and do not require compartmental replacement of the particles. The science of particulate contrast agents to evaluate tumor angiogenesis and anti-angiogenic treatment has critically been investigated. Particulate contrast agent of large size, i.e. microbubbles is used ultrasound imaging [15, 101].

3.3 Nano-phytomedicine Based Imaging of Permeability and Retention Through Biological Membrane and Site-Specific Drug Delivery

Solid tumors are normally distinguished by porous blood vessels, which facilitate the extravasations of nano-sized nanocarriers with a size range of 50–250 nm. The augmented permeability of tumor vasculature is the result of deregulated angiogenesis and higher release of vascular permeability increasing factors like VEGF. The disparity between the arrangement of new blood vessels and their maturation results to the formation of irregular endothelial layer that is distinguished by densely packed endothelium (size 300 nm). In addition to the leakiness of the tumor blood vessels, EPR also rely on the information that solid tumors be short of functional lymphatic drainage, owing to the dysfunctional lymph angiogenesis and compression of lymphatic vessels, that restricts the excretion of extravasated nanoparticle formulations from the intended site, leading to long-term retention in the pathological tissues [102]. Targeted drug delivery through nanoparticles is widely investigated. Poly N-(2-hydroxypropyl) methacrylamide (pHPMA)-copolymers have been applied as novel carrier systems for site-specific drug delivery. These gemcitabine and doxorubicin loaded copolymers were activated with gadolinium and 131I for imaging uses in addition to the therapeutic actions. MRI was used to assess biodistribution of the copolymers. Different types of copolymers have been radiolabeled with 131I for critical evaluation of tumor and organ retention of the polymeric systems intended for drug delivery. Tumor accumulation of co-polymer is increased when size of the nanocarrier radiolabeled with 131I is very high [103]. Soundararajan et al. investigated the therapeutic effectiveness of 186Re-labeled Doxilin in combination to the radiofrequency ablation therapy in HNSCC bearing rats. This study was done in terms of tumors volume and imaging of 18F-FDG PET. Results demonstrated that 186Re-labeled Doxil and radiofrequency ablation therapy combination reduces tumor enlargement, improves the formation of drug pool in tumors and improves treatment efficiency [104]. Koukourakis et al., showed utility of radiolabeled stealth liposome of doxorubicin that helped in the observation and measurement of EPR-mediated passive drug accumulation into tumors. This investigation demonstrated the ability of nanoparticles in the imaging of the pathways of the drug meant for site-specific delivery [105].

4 Future Prospects

Nano-sized herbal medicines can potentially enhance the biological activity and overcome the problems associated with pure herbal drugs. Nanotherapeutic products, which are more complex in structure and more expensive than conventional alternatives, are designed to provide an overall reduction in health care costs. This reduction in health care costs is likely to be obtained by increasing the nanotherapeutic efficacy, reducing the length of in-patient stay, reducing personal health care costs, and the effective treatment of expensive major diseases. The pharmaceutical regulatory environment, health care policies, demographics, and the wider economic environment directly affect the nanomedicine-related drug market. In the future, nanopharmaceuticals may alter the human body in the ways that we cannot imagine now but it is essential to consider benefits and side effects of the use of nanopharmaceuticals.

5 Conclusion

In last few decades, an array of nano-phytomedicines has been investigated as potential diagnostic and therapeutic agents. Extensive research is being carried out on novel drug delivery and site-specific delivery of plant actives and extracts for therapeutic purposes. However, investigation in this area is in the investigative phase. Several shortcomings associated with nano-phytomedicine are still needed to be addressed in research and development. Extra efforts should be made for investigating biomaterials based carriers for developing carriers of superior quality, which can improve the safety of the drug, improve activity, and overall quality and safety of the actives. Lipid solubility and molecular size are the prime points of concern for systemic absorption of drug molecules through biofilms after oral or topical administration. In spite of the excellent biological activity in vitro, plant extracts and their bio-actives exhibit few or no in vivo effects due to their poor fat solubility or inappropriate molecular size or both, resulting in poor absorption and bioavailability. Therefore, therapeutic potential of the herbs should be explored via value-added systems meant for drug delivery. When applied in the form of novel drug delivery system, standardized plant extracts or nano-phytoconstituents exhibit a better absorption characteristic, which allows them to pass through biofilms to enhance bioavailability. Diagnostic perspective calls for a significant requirement for a diagnostic tool capable enough for the EPR assessment. One of the efficient nanocarriers, i.e. theranostic nanoparticles have been utilized to visualize and measure biodistribution and accumulation of nanocarriers at the target site, to monitor drug release and long-term drug effectiveness, and to envisage a probable treatment response. Molecular imaging of mononuclear phagocytic systems, tumor formation, angiogenesis, and permeability and retention through biological membrane and targeted drug delivery by nanocarriers open new vista of NDDS application that have potential to change the current prospect of NDDS based research in near future. In the future, nanopharmaceuticals may alter the human body in the ways that we cannot imagine now but it is essential to consider benefits and side effects of the use of nanopharmaceuticals.

References

1. Javed MN, Alam MS, Waziri A, Pottoo FH, Yadav AK, Hasnain MS, Almalki FA (2019) QbD applications for the development of nanopharmaceutical products. In: Pharmaceutical quality by design. Academic, New York, pp 229–253
2. Javed MN, Kohli K, Amin S (2018) Risk assessment integrated QbD approach for development of optimized bicontinuous mucoadhesive limicubes for oral delivery of rosuvastatin. AAPS PharmSciTech 19(3):1377–1391
3. Mishra N, Yadav KS, Rai VK, Yadav NP (2017) Polysaccharide encrusted multilayered nanocolloidal system of andrographolide for improved hepatoprotection. AAPS PharmSciTech 18 (2):381–392
4. Alam MS, Javed MN, Pottoo FH, Waziri A, Almalki FA, Hasnain MS, Garg A, Saifullah MK (2019) QbD approached comparison of reaction mechanism in microwave synthesized gold nanoparticles and their superior catalytic role against hazardous nirto-dye. Appl Organomet Chem 33(9):e5071
5. Alam MS, Garg A, Pottoo FH, Saifullah MK, Tareq AI, Manzoor O, Mohsin M, Javed MN (2017) Gum ghatti mediated, one pot green synthesis of optimized gold nanoparticles: Investigation of process-variables impact using Box-Behnken based statistical design. Int J Biol Macromol 104:758–767
6. Hasnain MS, Javed MN, Alam MS, Rishishwar P, Rishishwar S, Ali S, Nayak AK, Beg S (2019) Purple heart plant leaves extract-mediated silver nanoparticle synthesis: optimization by Box-Behnken design. Mater Sci Eng 99:1105–1114
7. Sinha P, Srivastava N, Rai VK, Mishra R, Ajayakumar PV, Yadav NP (2019) A novel approach for dermal controlled release of salicylic acid for improved anti-inflammatory action: combination of hydrophilic-lipophilic balance and response surface methodology. J Drug Deliv Sci Technol 52:870–884
8. Sharma S, Javed MN, Pottoo FH, Rabbani SA, Barkat M, Sarafroz M, Amir M (2019) Bioresponse inspired nanomaterials for targeted drug and gene delivery. Pharm Nanotechnol 7(3):220–233
9. Mishra S, Sharma S, Javed MN, Pottoo FH, Barkat MA, Alam MS, Amir M, Sarafroz M (2019) Bioinspired nanocomposites: applications in disease diagnosis and treatment. Pharm Nanotechnol 7(3):206–219

10. Pottoo FH, Javed M, Barkat M, Alam M, Nowshehri JA, Alshayban DM, Ansari MA (2019) Estrogen and serotonin: complexity of interactions and implications for epileptic seizures and epileptogenesis. Curr Neuropharmacol 17(3):214–231
11. Pottoo FH, Tabassum N, Javed MN, Nigar S, Rasheed R, Khan A, Barkat MA, Alam MS, Maqbool A, Ansari MA, Barreto GE (2019) The synergistic effect of raloxifene, fluoxetine, and bromocriptine protects against pilocarpine-induced status epilepticus and temporal lobe epilepsy. Mol Neurobiol 56(2):1233–1247
12. Nigar S, Pottoo FH, Tabassum N, Verma SK, Javed MN (2016) Molecular insights into the role of inflammation and oxidative stress in epilepsy. J Adv Med Pharm Sci 10(1):1–9
13. Pottoo FH, Tabassum N, Javed MN, Nigar S, Sharma S, Barkat MA, Alam MS, Ansari MA, Barreto GE, Ashraf GM (2020) Raloxifene potentiates the effect of fluoxetine against maximal electroshock induced seizures in mice. Eur J Pharm Sci. https://doi.org/10.1016/j.ejps.2020.105261
14. Pottoo FH, Barkat MA, Ansari MA, Javed MN, Jamal QM, Kamal MA (2019) Nanotechnologoical based miRNA intervention in the therapeutic management of neuroblastoma. Sem Cancer Biol. https://doi.org/10.1016/j.semcancer.2019a.09.017
15. Baetke SCLT, Kiessling F (2015) Applications of nanoparticles for diagnosis and therapy of cancer. J Radiol 88:20150207
16. Rai VK, Mishra N, Yadav KS, Yadav NP (2018) Nanoemulsion as pharmaceutical carrier for dermal and transdermal drug delivery: formulation development, stability issues, basic considerations and applications. J Control Release 270:203–225
17. Sinha P, Srivastava S, Mishra N, Singh DK, Luqman S, Chanda D et al (2016) Development, optimization, and characterization of a novel tea tree oil nanogel using response surface methodology. Drug Dev Ind Pharm 42(9):1434–1445
18. Mishra N, Rai VK, Yadav KS, Sinha P, Kanaujia A, Chanda D et al (2016) Encapsulation of mentha oil in chitosan polymer matrix alleviates skin irritation. AAPS PharmSciTech 17 (2):482–492
19. Gunasekaran T, Haile T, Nigusse T, Dhanaraju MD (2014) Nanotechnology: an effective tool for enhancing bioavailability and bioactivity of phytomedicine. Asian Pac J Trop Biomed 4 (Suppl 1):S1–S7
20. Elmowafy M, Viitala T, Ibrahim HM, Abu-Elyazid SK, Samy A, Kassem A et al (2013) Silymarin loaded liposomes for hepatic targeting: in vitro evaluation and HepG2 drug uptake. Eur J Pharm Sci 50(2):161–171
21. Nanda S, Nanda A, Lohan S, Kaur R, Singh B (2016) Nanocosmetics: performance enhancement and safety assurance. In: Nanobiomaterials in galenic formulations and cosmetics. Elsevier, London, pp 47–67
22. Post-White J, Ladas EJ, Kelly K (2007) Advances in the use of milk thistle (Silybum marianum). Integr Cancer Ther 6(2):104–109
23. Polyak SJ, Morishima C, Lohmann V, Pal S, Lee DY, Liu Y et al (2010) Identification of hepatoprotective flavonolignans from silymarin. Proc Natl Acad Sci U S A 107 (13):5995–5999
24. Rashidinejad A, Birch EJ, Sun-Waterhouse D, Everett DW (2014) Delivery of green tea catechin and epigallocatechin gallate in liposomes incorporated into low-fat hard cheese. Food Chem 156:176–183
25. Pandita D, Ahuja A, Velpandian T, Lather V, Dutta T, Khar RJ (2009) Characterization and in vitro assessment of paclitaxel loaded lipid nanoparticles formulated using modified solvent injection technique. Pharmazie 64(5):301–310
26. Yan F, Li L, Deng Z, Jin Q, Chen J, Yang W et al (2013) Paclitaxel-liposome–microbubble complexes as ultrasound-triggered therapeutic drug delivery carriers. J Control Release 166 (3):246–255
27. Tian J, Wang L, Wang L, Ke X (2014) A wogonin-loaded glycyrrhetinic acid-modified liposome for hepatic targeting with anti-tumor effects. Drug Deliv 21(7):553–559

28. Kakkar V, Singh S, Singla D, Kaur IP (2011) Exploring solid lipid nanoparticles to enhance the oral bioavailability of curcumin. Mol Nutr Food Res 55(3):495–503
29. Liu Z, Okeke CI, Zhang L, Zhao H, Li J, Aggrey MO et al (2014) Mixed polyethylene glycol-modified breviscapine-loaded solid lipid nanoparticles for improved brain bioavailability: preparation, characterization, and in vivo cerebral microdialysis evaluation in adult Sprague Dawley rats. AAPS PharmSciTech 15(2):483–496
30. Boots AW, Haenen GR, Bast A (2008) Health effects of quercetin: from antioxidant to nutraceutical. Eur J Pharmacol 585(2-3):325–337
31. Capiau S, Stove VV, Lambert WE, Stove CP (2012) Prediction of the hematocrit of dried blood spots via potassium measurement on a routine clinical chemistry analyzer. Anal Chem 85(1):404–410
32. Si K, Wei L, Yu X, Wu F, Li X, Li C et al (2016) Effects of (+)-usnic acid and (+)-usnic acid-liposome on Toxoplasma gondii. Exp Parasitol 166:68–74
33. Cocco G, Chu DC, Pandolfi S (2010) Colchicine in clinical medicine. A guide for internists. Eur J Intern Med 21(6):503–508
34. Thomas OS, Weber W (2019) Overcoming physiological barriers to nanoparticle delivery-are we there yet? Front Bioeng Biotechnol 7:415
35. Marianecci C, Rinaldi F, Mastriota M, Pieretti S, Trapasso E, Paolino D et al (2012) Anti-inflammatory activity of novel ammonium glycyrrhizinate/niosomes delivery system: human and murine models. J Control Release 164(1):17–25
36. Yen F-L, Wu T-H, Lin L-T, Cham T-M, Lin C-C (2008) Nanoparticles formulation of Cuscuta chinensis prevents acetaminophen-induced hepatotoxicity in rats. Food Chem Toxicol 46 (5):1771–1777
37. Vilos C, Morales FA, Solar PA, Herrera NS, Gonzalez-Nilo FD, Aguayo DA et al (2013) Paclitaxel-PHBV nanoparticles and their toxicity to endometrial and primary ovarian cancer cells. Biomaterials 34(16):4098–4108
38. Min KH, Park K, Kim Y-S, Bae SM, Lee S, Jo HG et al (2008) Hydrophobically modified glycol chitosan nanoparticles-encapsulated camptothecin enhance the drug stability and tumor targeting in cancer therapy. J Control Release 127(3):208–218
39. Chen Y, Lin X, Park H, Greever R (2009) Study of artemisinin nanocapsules as anticancer drug delivery systems. Nanomedicine 5(3):316–322
40. Xue M, Yang M-X, Zhang W, Li X-M, Gao D-H, Ou Z-M et al (2013) Characterization, pharmacokinetics, and hypoglycemic effect of berberine loaded solid lipid nanoparticles. Int J Nanomedicine 8:4677–4687
41. Su JY, Tan LR, Lai P, Liang HC, Qin Z, Ye MR et al (2012) Experimental study on anti-inflammatory activity of a TCM recipe consisting of the supercritical fluid CO2 extract of Chrysanthemum indicum, patchouli oil and zedoary turmeric oil in vivo. J Ethnopharmacol 141(2):608–614
42. Chen L, Zhao X, Cai J, Guan Y, Wang S, Liu H et al (2013) Triptolide-loaded microemulsion-based hydrogels: physical properties and percutaneous permeability. Acta Pharm Sin 3 (3):185–192
43. Zhang C, Gu C, Peng F, Liu W, Wan J, Xu H et al (2013) Preparation and optimization of triptolide-loaded solid lipid nanoparticles for oral delivery with reduced gastric irritation. Molecules 18(11):13340–13356
44. Su Y, Fu Z, Zhang J, Wang W, Wang H, Wang Y et al (2008) Microencapsulation of Radix salvia miltiorrhiza nanoparticles by spray-drying. Powder Technol 184(1):114–121
45. Xu P, Yin Q, Shen J, Chen L, Yu H, Zhang Z et al (2013) Synergistic inhibition of breast cancer metastasis by silibinin-loaded lipid nanoparticles containing TPGS. Int J Pharm 454 (1):21–30
46. Bhattacharya SJ (2009) Phytosomes: the new technology for enhancement of bioavailability of botanicals and nutraceuticals. Int J Health Res 2(3):225–232
47. Pawar HA, Bhangale BD (2015) Phytosome as a novel biomedicine: a microencapsulated drug delivery system. J Bioanal Biomed 7(1):06–12

48. Kumar A, Kumar B, Singh S, Kaur B, Singh S (2017) A review on phytosomes: novel approach for herbal phytochemicals. Asian J Pharm Clin Res 10:41
49. Sharma S, Sahu AN (2016) Development, characterization, and evaluation of hepatoprotective effect of Abutilon indicum and piper longum phytosomes. Pharm Res 8(1):29–36
50. Nasrollahzadeh M, Sajadi SM (2015) Green synthesis of copper nanoparticles using Ginkgo biloba L. leaf extract and their catalytic activity for the Huisgen [3+2] cycloaddition of azides and alkynes at room temperature. J Colloid Interface Sci 457:141–147
51. Ju Ho P, Jun Sung J, Ki Cheon K, Jin TH (2018) Anti-inflammatory effect of Centella asiatica phytosome in a mouse model of phthalic anhydride-induced atopic dermatitis. Phytomedicine 43:110–119
52. Biedermann D, Vavříková E, Cvak L, Křen V (2014) Chemistry of silybin. Nat Prod Rep 31 (9):1138–1157
53. Flaig TW, Gustafson DL, Su L-J, Zirrolli JA, Crighton F, Harrison GS et al (2007) A phase I and pharmacokinetic study of silybin-phytosome in prostate cancer patients. Investig New Drugs 25(2):139–146
54. Singh A, Saharan VA, Singh M, Bhandari A (2011) Phytosome: drug delivery system for polyphenolic phytoconstituents. Iran J Pharm Sci 7(4):209–219
55. Yen FL, Wu TH, Lin LT, Cham TM, Lin CC (2009) Naringenin-loaded nanoparticles improve the physicochemical properties and the hepatoprotective effects of naringenin in orally-administered rats with CCl(4)-induced acute liver failure. Pharm Res 26(4):893–902
56. Nasrollahzadeh M, Sajadi SM (2015) Green synthesis of copper nanoparticles using Ginkgo biloba L. leaf extract and their catalytic activity for the Huisgen [3+ 2] cycloaddition of azides and alkynes at room temperature. Science 457:141–147
57. Jain N, Gupta BP, Thakur N, Jain R, Banweer J, Jain DK et al (2010) Phytosome: a novel drug delivery system for herbal medicine. Global J Pharm Res 2(4):224–228
58. Karimi N, Ghanbarzadeh B, Hamishehkar H, Pezeshki A, Mostafayi H, Gholian M (2015) Phytosome as novel delivery system for nutraceutical materials. Int J Curr Microbiol App Sci 4 (6):152–159
59. Barkat A, Khan B, Naveed A, Muhammad H, Khan HM, Waseem K et al (2011) Basics of pharmaceutical emulsions: a review. Afr J Pharm Pharmacol 525:2715–2725
60. Yin Y-M, Cui F-D, Mu C-F, Choi M-K, Kim JS, Chung S-J et al (2009) Docetaxel microemulsion for enhanced oral bioavailability: preparation and in vitro and in vivo evaluation. J Control Release 140(2):86–94
61. Wang X, Jiang Y, Wang Y-W, Huang M-T, Ho C-T, Huang Q (2008) Enhancing anti-inflammation activity of curcumin through O/W nanoemulsion. Food Chem 108:419–424
62. Megha MA, Unnma U, Rameshpathy M, Karikalan K, Sundaram V, Kumar V et al (2013) Formulation of nano-encapsulated poly-herbal ointment for antiinflammation. Pharm Lett 5:164–170
63. Zhao Y, Wang C, Chow AHL, Ren K, Gong T, Zhang Z et al (2010) Self-nanoemulsifying drug delivery system (SNEDDS) for oral delivery of Zedoary essential oil: formulation and bioavailability studies. Int J Pharm 383(1-2):170–177
64. Okur NÜ, Apaydın Ş, Yavaşoğlu NÜK, Yavaşoğlu A, Karasulu H (2011) Evaluation of skin permeation and anti-inflammatory and analgesic effects of new naproxen microemulsion formulations. Int J Pharm 416(1):136–144
65. Brahmer J, Reckamp KL, Baas P, Crinò L, Eberhardt WE, Poddubskaya E et al (2015) Nivolumab versus docetaxel in advanced squamous-cell non–small-cell lung cancer. N Engl J Med 373(2):123–135
66. Reyes-Escogido M, Gonzalez-Mondragon EG, Vazquez-Tzompantzi EJM (2011) Chemical and pharmacological aspects of capsaicin. Molecules 16(2):1253–1270
67. Biedermann D, Vavrikova E, Cvak L, Kren V (2014) Chemistry of silybin. Nat Prod Rep 31 (9):1138–1157

68. Ascenso A, Raposo S, Batista C, Cardoso P, Mendes T, Praça FG et al (2015) Development, characterization, and skin delivery studies of related ultradeformable vesicles: transfersomes, ethosomes, and transethosomes. Int J Nanomedicine 10:5837
69. More MS, Shende MA, Kolhe DB, Jaiswal N (2012) Herbosomes: herbo-phospholipid complex an approach for absorption enhancement. Chemistry 3(8):946–955
70. Cabrera C, Artacho R, Giménez R (2006) Beneficial effects of green tea--a review. J Am Coll Nutr 25(2):79–99
71. Paolino D, Lucania G, Mardente D, Alhaique F, Fresta M (2005) Ethosomes for skin delivery of ammonium glycyrrhizinate: in vitro percutaneous permeation through human skin and in vivo anti-inflammatory activity on human volunteers. J Control Release 106:99–110
72. Chen J-G, Liu Y-F, Gao T (2010) Preparation and anti-inflammatory activity of triptolide ethosomes in an erythema model. J Liposome Res 20:297–303
73. Mishra D, Panda G, Kumar P, Singh S (2017) Novel drug delivery system for herbal formulation in cancer treatment. World J Pharm Res 6:341–353
74. Nirved U, Lokesh V, Mishra G, Hemant J (2012) Formulation and evaluation of ethosomes of Sesbania grandiflora Linn. Seeds. Novel Sci Int J Pharm Sci 1:1
75. Singh P, Prakash DEV, Ramesh B, Singh N, Mani T (2011) Biodegradable polymeric microspheres as drug carriers: a review. Indian J Novel Drug Deliv 2:70–82
76. Geninatti Crich S, Bussolati B, Tei L, Grange C, Esposito G, Lanzardo S et al (2006) Magnetic resonance visualization of tumor angiogenesis by targeting neural cell adhesion molecules with the highly sensitive gadolinium-loaded apoferritin probe. Cancer Res 66(18):9196–9201
77. Zhang C, Jugold M, Woenne EC, Lammers T, Morgenstern B, Mueller MM et al (2007) Specific targeting of tumor angiogenesis by RGD-conjugated ultrasmall superparamagnetic iron oxide particles using a clinical 1.5-T magnetic resonance scanner. Cancer Res 67 (4):1555–1562
78. Kiessling F, Morgenstern B, Zhang C (2007) Contrast agents and applications to assess tumor angiogenesis in vivo by magnetic resonance imaging. Curr Med Chem 14(1):77–91
79. Winter PM, Cai K, Chen J, Adair CR, Kiefer GE, Athey PS et al (2006) Targeted PARACEST nanoparticle contrast agent for the detection of fibrin. Magn Reson Med 56(6):1384–1388
80. Cassidy MC, Chan HR, Ross BD, Bhattacharya PK, Marcus CM (2013) In vivo magnetic resonance imaging of hyperpolarized silicon particles. Nat Nanotechnol 8:363
81. Leike JSA, Ehritt C, Krause W (1996) Biodistribution and CT-imaging characteristics of iopromide-carrying liposomes in rats. J Liposome Res 6:665–680
82. Torchilin VP, Frank-Kamenetsky MD, Wolf GL (1999) CT visualization of blood pool in rats by using long-circulating, iodine-containing micelles. Acad Radiol 6(1):61–65
83. Liu Y (2018) Hybrid BaYbF5 nanoparticles: novel binary contrast agent for high-resolution in vivo X-ray computed tomography angiography. In: Liu Y (ed) Multifunctional nanoprobes: from design validation to biomedical applications. Springer, Singapore, pp 105–120
84. Cho EC, Glaus C, Chen J, Welch MJ, Xia Y (2010) Inorganic nanoparticle-based contrast agents for molecular imaging. Trends Mol Med 16(12):561–573
85. Rabin O, Manuel Perez J, Grimm J, Wojtkiewicz G, Weissleder R (2006) An X-ray computed tomography imaging agent based on long-circulating bismuth sulphide nanoparticles. Nat Mater 5(2):118–122
86. Kiessling F, Bzyl J, Fokong S, Siepmann M, Schmitz G, Palmowski M (2012) Targeted ultrasound imaging of cancer: an emerging technology on its way to clinics. Curr Pharm Des 18(15):2184–2199
87. Gao Z, Kennedy AM, Christensen DA, Rapoport NY (2008) Drug-loaded nano/microbubbles for combining ultrasonography and targeted chemotherapy. Ultrasonics 48(4):260–270
88. Kim K, Kim JH, Park H, Kim YS, Park K, Nam H et al (2010) Tumor-homing multifunctional nanoparticles for cancer theragnosis: simultaneous diagnosis, drug delivery, and therapeutic monitoring. J Control Release 146(2):219–227
89. Gao X, Cui Y, Levenson RM, Chung LW, Nie S (2004) In vivo cancer targeting and imaging with semiconductor quantum dots. Nat Biotechnol 22(8):969–976

90. Santra S, Dutta D, Walter GA, Moudgil BM (2005) Fluorescent nanoparticle probes for cancer imaging. Technol Cancer Res Treat 4(6):593–602
91. Cai X, Li W, Kim CH, Yuan Y, Wang LV, Xia Y (2011) In vivo quantitative evaluation of the transport kinetics of gold nanocages in a lymphatic system by noninvasive photoacoustic tomography. ACS Nano 5(12):9658–9667
92. De la Zerda A, Zavaleta C, Keren S, Vaithilingam S, Bodapati S, Liu Z et al (2008) Carbon nanotubes as photoacoustic molecular imaging agents in living mice. Nat Nanotechnol 3 (9):557–562
93. Akers WJ, Kim C, Berezin M, Guo K, Fuhrhop R, Lanza GM et al (2011) Noninvasive photoacoustic and fluorescence sentinel lymph node identification using dye-loaded perfluorocarbon nanoparticles. ACS Nano 5(1):173–182
94. Xie H, Wang ZJ, Bao A, Goins B, Phillips WT (2010) In vivo PET imaging and biodistribution of radiolabeled gold nanoshells in rats with tumor xenografts. Int J Pharm 395(1-2):324–330
95. Majmudar MD, Yoo J, Keliher EJ, Truelove JJ, Iwamoto Y, Sena B et al (2013) Polymeric nanoparticle PET/MR imaging allows macrophage detection in atherosclerotic plaques. Circ Res 112(5):755–761
96. Ocampo-Garcia BE, Ramirez Fde M, Ferro-Flores G, De Leon-Rodriguez LM, Santos-Cuevas CL, Morales-Avila E et al (2011) (99m)Tc-labelled gold nanoparticles capped with HYNIC-peptide/mannose for sentinel lymph node detection. Nucl Med Biol 38(1):1–11
97. Koukourakis MI, Koukouraki S, Giatromanolaki A, Kakolyris S, Georgoulias V, Velidaki A et al (2000) High intratumoral accumulation of stealth liposomal doxorubicin in sarcomas--rationale for combination with radiotherapy. Acta Oncol 39(2):207–211
98. Brouwer OR, Buckle T, Vermeeren L, Klop WM, Balm AJ, van der Poel HG et al (2012) Comparing the hybrid fluorescent-radioactive tracer indocyanine green-99mTc-nanocolloid with 99mTc-nanocolloid for sentinel node identification: a validation study using lymphoscintigraphy and SPECT/CT. J Nucl Med 53(7):1034–1040
99. Pattni BS, Chupin VV, Torchilin VP (2015) New developments in liposomal drug delivery. Chem Rev 115(19):10938–10966
100. Weissleder R, Elizondo G, Stark DD, Hahn PF, Marfil J, Gonzalez JF et al (1989) The diagnosis of splenic lymphoma by MR imaging: value of superparamagnetic iron oxide. Am J Roentgenol 152(1):175–180
101. Palmowski M, Huppert J, Ladewig G, Hauff P, Reinhardt M, Mueller MM et al (2008) Molecular profiling of angiogenesis with targeted ultrasound imaging: early assessment of antiangiogenic therapy effects. Mol Cancer Ther 7(1):101–109
102. Baetke SC, Lammers T, Kiessling F (2015) Applications of nanoparticles for diagnosis and therapy of cancer. Br J Radiol 88(1054):20150207
103. Lammers T, Subr V, Peschke P, Kühnlein R, Hennink WE, Ulbrich K et al (2008) Image-guided and passively tumour-targeted polymeric nanomedicines for radiochemotherapy. Br J Cancer 99(6):900–910
104. Soundararajan A, Bao A, Phillips WT, Perez R, Goins BA (2009) [186Re]Liposomal doxorubicin (Doxil): in vitro stability, pharmacokinetics, imaging and biodistribution in a head and neck squamous cell carcinoma xenograft model. Nucl Med Biol 36(5):515–524
105. Koukourakis Sofia Koukouraki A, Nikolaos N, Karkavitsas MI (2000) High intratumoral accumulation of stealth liposomal doxorubicin in sarcomas: rationale for combination with radiotherapy. Acta Oncol 39(2):207–211

The manufacturer's authorised representative in the EU is Springer Nature Customer Service Centre GmbH, Europaplatz 3, 69115 Heidelberg, Germany. If you have any concerns regarding our products, please contact ProductSafety@springernature.com

Printed and bound by CPI Group (UK) Ltd, Croydon, CR0 4YY
30/09/2026
02225559-0007